ANIMAL AND HUMAN HEALTH AND WELFARE: A COMPARATIVE PHILOSOPHICAL ANALYSIS

ANIMAL AND HUMAN HEALTH AND WELFARE: A COMPARATIVE PHILOSOPHICAL ANALYSIS

Lennart Nordenfelt

Department of Health and Society
Linköping University
SE-58183 Linköping
Sweden

www.cabi.org

CABI is a trading name of CAB International

CABI Head Office	CABI North American Office
Nosworthy Way	875 Massachusetts Avenue
Wallingford	7th Floor
Oxfordshire OX10 8DE	Cambridge, MA 02139
UK	USA
Tel: +44 (0)1491 832111	Tel: +1 617 395 4056
Fax: +44 (0)1491 833508	Fax: +1 617 354 6875
E-mail: cabi@cabi.org	E-mail: cabi-nao@cabi.org
Website: www.cabi.org	

A catalogue record for this book is available from the British Library, London, UK.

Library of Congress Cataloging-in-Publication Data

Nordenfelt, Lennart. 1945-

Animal and human health and welfare : a comparative philosophical analysis / Lennart Nordenfelt.

　　　　p. cm.
　　　Includes bibliographical references and index.
　　　ISBN-13: 978-1-84593-059-2 (alk. paper)
　　　ISBN-10: 1-84593-059-2 (alk. paper)
　　1. Health—Philosophy. 2. Medicine—Philosophy. 3. Veterinary medicine—Philosophy. I. Title.
　　　　[DNLM: 1. Philosophy, Medical. 2. Health. 3. Quality of Life.
4. Animal Welfare. 5. Animal Husbandry.　W 61 N829a 2006]
R723.N6575 2006
610'.1--dc22
　　　2005021603

ISBN-10:　　　1-84593-059-2
ISBN-13:　　978-1-84593-059-2

Typeset by MRM Graphics Ltd, Winslow, Bucks
Printed and bound in the UK by Biddles, King's Lynn

Contents

Preface

This work is one of the results of a research project called 'Health and Welfare in the Worlds of Humans and Animals' that has been financed by the Swedish Council for Working Life and Social Research, and which has been running since 2002 and will be concluded by the end of 2005. The participants of this project have been, besides myself, the following: Professor Bo Algers and Associate Professor Stefan Gunnarsson, Swedish University of Agricultural Sciences; Associate Professor Ingemar Lindahl, Department of Philosophy, Stockholm University; Professor Anders Nordgren, Centre for Applied Ethics, Linköping University; and Mr Henrik Lerner, doctoral student at the Department of Health and Society, Linköping University. Thanks are due to these colleagues who have made many valuable comments regarding the subject matter of this book during the course of the project.

I am also very grateful to Professors David Fraser, University of British Columbia, and Donald Broom, University of Cambridge, for their very detailed and helpful comments on earlier versions of this manuscript. They have generously shared with me their great experience from animal studies and their profound insights into the theory of animal welfare. Without their help I would have made several embarrassing mistakes. Needless to say, I myself am responsible for the remaining errors.

Since I have previously written extensively in the area of human health and quality of life I have been able to use some arguments and even a few complete passages from works of mine that have been published earlier. In three cases, to be particularly mentioned here, I have made substantial quotations from my own works. This concerns, first, Chapter 1, 'An overview of historical conceptions of human health'. In this case Sage Press, Thousand Oaks, California, has permitted me to use material from my article on health published in G.A. Albrecht, J. Bichenbach, D. Mitchell, W.O.Schalick and S. Snyder (eds) *World Encyclopedia of Disability* (2005). Second, with regard to Chapter 3, 'The place of evolutionary theory in the philosophy of health and welfare', I have received permission from Rodopi Publishers, Amsterdam, to use passages from my article: 'Health as natural function', in L. Nordenfelt and P.-E. Liss (eds) *Dimensions of Health and Health Promotion* (2003). Third, for my presentation of 'Two classic theories of human welfare' in Chapter 4, I have used some passages from my book *Quality of Life, Health and Happiness* (1993), Avebury, Aldershot. Since this is now out of print all the rights to its contents belong to me. I am also grateful to the Universities Federation for Animal Welfare for letting me use a figure (Fig. 17.1) that appeared in Fraser, D., Weary, D.M., Pajor, E.A. and Milligan, B.N. (1997) A scientific conception of animal welfare that reflects ethical concerns. *Animal Welfare* 6, 187–205.

More complete versions of my holistic theory of health and of my happiness theory of welfare can be found in my *On the Nature of Health* (1987/1995), Kluwer Publishers, Dordrecht, and *Quality of Life, Health and Happiness* (1993), Avebury. Some revisions of these theories appear in *Action, Ability and Health* (2000), Kluwer Publishers.

Lennart Nordenfelt
Linköping, August 2005

Introduction

For about 30 years there has been an extensive discussion within philosophy (as well as theoretical anthropology, sociology and psychology) concerning the concept of health and related concepts in medicine and other health sciences. Examples of such concepts are, on the positive side, fitness and quality of life, and, on the negative side, disease, illness, sickness, disability and handicap. The discussion has been pursued mainly within two philosophical sub-disciplines, philosophy of medical science and ethics. The purposes, accordingly, have been different. Within the philosophy of science discourse the debate has been focused on the notion of disease and the aim has been to find universal criteria for the identification of states as diseases. Within the ethics discourse the focus has often (but not exclusively) been on the concept of health and on health as a goal of medicine and health care. A crucial ethical concern has also been to find relevant criteria of mental illness, not least in the context of forensic psychiatry. The different purposes have been reflected in the final proposals for the characterization of the concepts. There have also, however, been serious attempts to combine the scientific and ethical concerns.

Also in veterinary science and animal science there is a continuing theoretical discussion with regard to basic concepts but it has had a slightly different character and has to a great extent had other purposes. The animal science analysis has focused on the notion of welfare instead of the notions of health and quality of life. Moreover, the motivation for this analysis has been largely ethical. The analysis has issued from the lively ethical debate since the 1960s about our treatment of animals both in the food industry and in medical research. The starting-point for this discussion was the publication of Ruth Harrison's *Animal Machines* (1964), where she particularly criticized modern farming methods. As a result of Harrison's book and the subsequent discussion a UK government committee (the Brambell Committee) was asked to 'examine the conditions in which livestock are kept under systems of intensive husbandry and to advise whether standards ought to be set in the interest of their welfare and if so what they should be'.

The Brambell Committee investigated the notion of animal welfare thoroughly and they rejected as oversimplified and incomplete the view that an animal's productivity could be taken as decisive evidence of the animal's state of welfare. They attached importance to animal health and also to the animal's being able to engage in its natural behaviour. The advantage of providing an animal with shelter and protection must be weighed against the disadvantage of inhibiting its ability to move as it has been wont (or even designed) to do.

In the review presented by the Brambell Committee (1965) they proposed that all farm animals should at least have the freedom to 'stand up, lie down, turn around, groom themselves and stretch their limbs'. These requirements were referred to as the five freedoms. The veterinarian John Webster, who was for a long time working on the UK Farm Animal Welfare Council (FAWC), found this list of animal freedoms too restricted. The list concentrated exclusively on space requirements and Webster sought for a list that covered all essential aspects of animal welfare. In the end he presented a more general list, where he, however, retained the idea of 'five freedoms'. This list was finally codified by the FAWC (1993), and has since become a standard for further discussions on animal welfare. (See Webster, 1994.) It reads as follows:

1. Freedom from thirst, hunger and malnutrition – by providing ready access to fresh water and a diet to maintain full health and vigour.
2. Freedom from discomfort – by providing a suitable environment including shelter and a comfortable resting area.
3. Freedom from pain, injury and disease – by prevention or rapid diagnosis and treatment.
4. Freedom to express normal behaviour – by providing sufficient space, proper facilities and company of the animal's own kind.
5. Freedom from fear and distress – by ensuring conditions which prevent mental suffering.

Following the Brambell report and the subsequent work of the FAWC there has been extensive popular and scientific discussion of the notion of animal welfare and its interpretation. Significant voices in the scientific discussion have been D.M. Broom, I.J.H. Duncan, M.S. Dawkins and D. Fraser, among others. Philosophers such as B.E. Rollin and P. Sandøe have specialized in the area and have diligently contributed, mainly to the ethical discussion.

It is striking, given the intensity and complexity of these discussions, that there has been little (indeed virtually no) communication between the human and animal science approaches to the concepts of health, quality of life and welfare. It is timely, therefore, to make a comparative analysis of the two areas of research.

The present book will attempt to make such a comparison. I will first present as a background some major theories of human health and welfare. Among these are the biostatistical theory and a few so-called holistic theories of health. On the quality-of-life side I present some classical theories (Aristotle and Bentham) and a number of contemporary theories (both objectivist and subjectivist ones). Upon this follows a more detailed discussion of some theories of animal welfare and to some extent health. (The literature that focuses on the *concept* of animal health is very limited.) I analyse coping theories, feeling and preference theories as well as complex theories of welfare. In the last, more constructive, part of the book I wish to test a comprehensive conceptual framework, first designed for the case of human health and welfare, on the animal field. The concept of health proposed is of a holistic kind, focusing on the individual's ability to achieve vital goals. The concept of welfare (or quality of life) is related to the global notion of satisfaction or happiness.

A central question to be put in an analysis of animal health and welfare is the following: what animals are we talking about? Does the study deal with all animals from primates to amoebas, or does it concern only such higher animals as play a role in relation to humans and are of interest from a human perspective? It is obvious that most theories of animal welfare deal with a limited set of animals. Central are livestock and pets. Sometimes animals in circuses and zoos are included, and on a few occasions wild animals. In the latter case the concern is still from a human perspective, i.e. the theorist discusses such wild animals as are or could be under some human control. However, a few theories in the area, such as Donald Broom's theory of welfare, claim to be valid for all species of animals.

Thus, this text occasionally applies to all animals, but mostly to a much more limited set of animals which are in various ways close to humans. The constructive part of my thesis very clearly focuses on higher animals to which some cognition and emotion can reasonably be attributed. This, however, is hardly any limitation in relation to the mainstream of the current discussion on animal health and welfare.

I will not attempt to draw a sharp border between such species as are relevant for my analysis and such as fall outside. First, my biological knowledge is too limited for that task. Second, I think there is little point in drawing such a sharp line. If my reasoning happens to be totally relevant for a certain set of higher animals, there might be many species in the animal world for which my reasoning could be at least partly relevant. Birds, for instance, are not within my focus. However, I will allow myself to take a few examples from the area of ornithology.

A few words about the composition of this book. Although the book deals with both humans and animals, and the main purpose is to make a comparison between notions in human medicine and animal science, there will be an emphasis on the notions of health and welfare with regard to animals. This will be quite evident not only in terms of the space dedicated to the disciplines but also in the fact that animal theories will here be put under more detailed scrutiny. There are two reasons for this imbalance. First, I have myself worked for a long time within philosophy of human medicine and health care and made quite close examinations of various theories of health and welfare (1987/1995, 1994, 2001 and 2003). There is no point in repeating these critical analyses in any detail here. I have therefore limited myself to presenting some of the standpoints on human health and welfare that are necessary for a comparison with the animal welfare discussion. Second, there exist far fewer critical studies of animal health and welfare theories. My presentation could therefore perhaps play some role as an original analysis of such theories.

Let me add a point on a terminological issue. As I will note during the discussion in this book, there is a lot of confusion about the terminology to be used. In the literature about human welfare the expression 'quality of life' is currently in fashion. This expression has also entered the animal welfare literature but has not gained much recognition (McMillan, 2000). In the animal science and the animal welfare literature the term 'welfare' is the most common. In the scientific literature on humans, in particular in sociology, 'welfare' often refers to socio-economic states of affairs. A society is a welfare society if social institutions are well-developed and if the standard of living is reasonably high. In several American animal science contributions also 'well-being' exists. It is then mostly used as a synonym of (European) 'welfare'.

In a few languages, including Swedish and German, and possibly English, one can trace an interesting lexical distinction between 'welfare' and 'well-being' (Swedish *välfärd* and *välbefinnande*, German *Wohlfahrt* and *Wohlbefinden*). In the ordinary usage of these languages the first of the pair of terms is more used to refer to external facts such as socio-economic conditions, whereas the second typically refers to subjective experiences. I have earlier (1993) used this lexical fact as a reason for introducing a technical distinction between welfare (external positive/negative facts) and well-being (positive/negative feelings) in my analysis of human quality of life. Elements of that analysis are included also in this book (see Part III). However, on the advice given by some prominent animal welfare scientists, I have not retained this terminological distinction. In the animal welfare context it would perhaps create more confusion than illumination to distinguish between welfare and well-being. I will therefore only use the term 'welfare' in the animal context (except, of course, when I am quoting or referring to authors using the term 'well-being') but sometimes talk about 'external welfare' when referring to such factors regarding an animal as contribute to its (inner) welfare.

I

Some Theories of Human Health and Welfare

1 An Overview of Historical Conceptions of Human Health

Introduction

The concept of human health is the most central one in medicine and in the health sciences in general. Health is, indeed, the foremost goal of medicine. Health also has a prominent position in many life contexts, is a crucial condition for pursuing a profession, for enjoying leisure activities, and indeed for living a good life in general. For example, health is a formal prerequisite for performing certain tasks or taking up certain occupations, such as that of soldier, police officer or firefighter. More compelling is the place of mental health as a condition for moral and criminal culpability.

It is significant that, in modern secular society, health has gained an extremely high position in many people's value hierarchies. A Swedish group of researchers (Kallenberg *et al.*, 1997) asked a representative sample of Swedes what are the highest values in their lives. A vast majority of these people put health at the top of their list, which also contained such values as wealth, high social status and good family relations. This social judgement can be contrasted with the ancient Platonian more restricted evaluation of health. According to Plato, it is an unsound condition in a society when people concentrate on their health and want to consult a physician anytime and about any question (*The Republic*; for a modern edition see Plato, 1998).

Etymologically, health is connected with the idea of wholeness. This is evident in the verb 'heal', with the sense of regaining wholeness. The healthy person is a person who is whole in the sense of having all the properties that should pertain to a human being. Health has thus traditionally been viewed as an ideal notion, a notion of perfection that very few people, if any, can completely attain. Today health also sometimes functions as an ideal notion. This is, indeed, the case with the formulation of health by the World Health Organization (WHO) in its initial declaration, published in 1948: 'Health is a state of complete physical, mental and social well-being and not only the absence of disease or injury'.

The notion of health is the object of scientific study from several points of view and within several disciplines. Besides research by those in medicine, public health, nursing and other paramedical disciplines, other investigations are based in anthropology, psychology, sociology and philosophy. In some of these disciplines the focus is on a particular aspect of the notion: in psychology, the experience of health and illness; in anthropology and sociology, health and illness as factors of social importance. Philosophical analyses of health have often involved an attempt to formulate global definitions of the idea. Thus, in the following, many references will be taken from philosophical theories of health.

The Varieties of Health

Health, thus, is a notion primarily applicable to a human being as a whole. On the other hand, there are more specific derivative notions. Ever since antiquity, and reinforced by the Cartesian distinction between body and mind, it has been natural to separate somatic health from mental health. The interpretations of mental health have varied over time. The ancient notion of mental health was closely connected to morality, whereby the mentally healthy person was a person who lived a virtuous life, but this idea has lost most, though not all, of its significance today. The idea of spiritual health is also current in the health sciences although it is not systematically recognized. Bernhard Häring (1987) is a leading spokesman for a notion of health including a spiritual dimension: 'A comprehensive understanding of human health includes the greatest possible harmony of all of man's forces and energies, the greatest possible spiritualization of man's bodily aspect and the finest embodiment of the spiritual' (p. 154).

The various categories of health have connections to each other. Sometimes bodily health has been given priority in the sense that it has been viewed as a prerequisite for mental health. Galen (c. 129–216/7) in some of his writings attempted to explain mental properties of the person in terms of specific mixtures of the bodily parts (Galen, 1997). Consider also the ancient proverb: *mens sana in corpore sano* (a healthy mind in a healthy body). In the modern discussion about mental illness, one position, favoured in particular by doctors, is that all mental illness has a somatic background, i.e. that all mental illnesses – if they exist at all – are basically somatic diseases (Szasz, 1974). The customary view, however, also in Western medicine, is that a person can at the same time be somatically healthy and mentally ill, or vice versa.

Health as Absence of Disease: the Idea of a Natural Function

Although health is often described in non-medical terms and with reference to non-medical contexts, it has its primary place and function as a medical concept. Health in the medical arena is contrasted in particular with disease, but also with injury, defect and disability. Culver and Gert (1982) have chosen the term 'malady' to cover the negative antipodes of health. In many medical contexts and in some philosophical reconstructions of the notion of health (Boorse, 1977, 1997), health has been defined as the absence of disease or the absence of malady. The perfectly healthy person therefore is the person who does not have any diseases or maladies.

If one looks upon the relationship between the concepts in this way, the burden of definition lies on the negative notions. Christopher Boorse (1997, p. 8), for instance, defines disease in the following terms: 'A disease is a type of internal state which is either an impairment of normal functional ability, i.e. a reduction of one or more functional abilities below typical efficiency, or a limitation on functional ability caused by environmental agents'. The notion of functional ability, in this theory, is in its turn related to the person's survival and reproduction, viz. his or her fitness. From this analysis it follows that we need not use the notion of disease in order to define health. The same idea can be formulated in the following positive terms: a person is completely healthy if, and only if, all his or her organs function with at least typical efficiency (in relation to survival and reproduction).

This idea of natural function is similar to, but not identical with, the one proposed by Jerome Wakefield (1992a and later), where the platform for analysis is biological evolutionary theory. The natural function F of an organ is, according to this idea, the function for which it has been designed through evolution. This means that the species in question (for instance, the human being) has been able to reproduce through history with the genetic set-up for the function F. This idea has been criticized partly because it relates the idea of health in the present context to developments in the past.

Health, Disease and Illness

In many contributions to the theory of health a distinction is made between the concepts of disease and illness (Fulford, 1989; Twaddle, 1993). The general idea behind this distinction – although it has been made in different ways by different authors – is that a disease is a deranged process in the person's body whereas an illness is the person's negative experiences, for instance pain or anguish, as a result of the disease. In addition, some theories include disability in illness (see below). The distinction between disease and illness has proved useful in several contexts, including the clinical one (Hellström, 1993), for separating the disease as a pathological phenomenon from its impact on the person as a whole. (For a criticism of the distinction between disease and illness, see Sundström, 1987.)

Health as Balance

An extremely powerful idea in the history of medicine is the one that health is constituted by bodily and mental balance. The healthy person is a person in balance, normally meaning that different parts and different functions of the human body and mind interlock harmoniously and keep each other in check. The Hippocratic and Galenic schools (Hippocrates, 460–380 BC and Galen, AD 129–216/7) were the first Western schools to develop this idea in a sophisticated way. They stated that a healthy body is one where the primary properties (wet, dry, cold, hot) of the body balance each other. In the medieval schools, following Galen, this idea was popularized and formulated in terms of a balance between the four bodily humours: blood, phlegm, yellow bile and black bile.

The idea of balance is strong in several non-Western medical traditions. The Ayurveda tradition in India, for instance, declares that there are three humours acting in the body, the breath (*vata*), the bile (*pitta*) and the phlegm (*kapha*). The proportions of the three humours vary from person to person, and their actions vary according to the season, the environment, the lifestyle of the individual, and his or her diet. In good health the humours are in equilibrium. Disease is the result of their imbalance (Singhal and Patterson, 1993).

Balance is a powerful idea also in modern Western thought, in particular within physiology. The idea is often to be recognized under the label of *homeostasis* (the Greek word for balance). Walter Cannon's (1871–1945) classical work on homeostasis (1932) describes in detail how the various physiological functions of the body control each other and interact in feedback loops in order to prevent major disturbances.

The idea of balance or *equilibrium* (the Latin word for balance) has a rather different interpretation in the writings of Ingmar Pörn (1993). Here balance is a concept pertaining to the relationship between a person's abilities and his or her goals. The healthy person, according to Pörn, is the person who can realize his or her goals and thus retain a balance between abilities and goals. (Cf. health as ability, below.)

Health as Well-being

It is an important aspect of health that the body and mind are well, both in order and function. But we may ask for the criteria of such well-functioning. How do we know that the body and mind function well? When is the body in balance?

A traditional answer is that the person's subjective well-being is the ultimate criterion. Simply put: when a person feels well, then he or she is healthy. This statement certainly entails problems, since a person can feel well and still have a serious disease in its initial stage. The general idea can, however, be modified to cover this case too. The individual with a serious disease will sooner or later have negative experiences such as pain, fatigue or anguish. Thus, the ultimate criterion of a person's health is his or her present or future well-being. (For a different approach suggesting that complete health is compatible with the existence of disease, see Nordenfelt, 1987/1995, 2000.)

It is a difficult task to characterize the well-being constituting health. If one includes too much in the concept there is a risk of identifying health with happiness. It is, indeed, a common accusation directed against the WHO definition that it falls into this trap. Many critics say health cannot reasonably be identical with complete physical, mental and social well-being. Taken to its absurd conclusion, this conception could imply that all people who are not completely successful in life are to be deemed unhealthy.

Some authors (Leder, 1990; Gadamer, 1993) have pointed out that phenomenological health (or health as experienced) tends to remain as a forgotten background. Health is in daily life hardly recognized at all by its subjects. People are reminded of their previous health only when it is disrupted, when they experience the pain, nausea or anguish of illness. Health is 'felt' only under special circumstances, the major instance being after periods of illness when the person experiences relief in contrast to the previous suffering.

Thus, although well-being or absence of ill-being is an important trait in health, most modern positive characterizations of health have focused on other traits. One such trait is health as a condition for action, i.e. ability.

Health as Ability

A number of authors in modern philosophy of health have emphasized the place of health as a foundation for achievement (Parsons, 1972; Whitbeck, 1981; Seedhouse, 1986; Nordenfelt, 1987/1995; Fulford, 1989). In fact they argue, in partly different ways, that the dimension ability/disability is the core dimension determining whether health or ill-health is the case. A healthy person has the ability to do what he or she needs to do, and the unhealthy person is prevented from performing one or more of these actions. There is a connection between this conception and the one that illness entails suffering. Disability is often the result of feelings such as pain, fatigue or nausea.

The formidable task for these theorists is to characterize the set of actions that a healthy person should be able to perform. Parsons and Whitbeck refer to the subject's wants, i.e. the healthy person's being able to do what he or she wants, Seedhouse to the person's conscious choices, and Fulford to such actions as could be classified as 'ordinary doings'. I myself settle for what I call the subject's vital goals. These goals need not be consciously chosen (also babies and people with dementia have vital goals). The goals have the status as vital goals because they are states of being that are necessary conditions for the person's happiness in the long run. Health in my theory is thus conceptually related to, but not identical with, happiness.

Although it is evident that health, as ordinarily understood, is connected with ability, and ill-health with disability, one may still doubt whether the dimension ability/disability can remain the sole criterion of health/ill-health. An important argument concerns those disabled people who are not ill, according to common understanding, and who do not consider themselves to be ill. These people are to be classified as unhealthy according to the ability theories of health.

One way out is to say that disabled people (given that their disability is assessed in relation to their individual vital goals) are all unhealthy. However,

they are not all ill and they do not all have diseases. Another way out, proposed by Fredrik Svenaeus (2001), is to say that there is a phenomenological
difference between the disabled unhealthy person and the disabled healthy
person. The unhealthy person has a feeling of not being 'at home', with regard
to his or her present state of body or mind. This feeling is not present in the
case of the disabled in general.

Health and Value. Naturalists and Normativists

A crucial theoretical problem in the characterization of health is whether this
notion is in the old positivistic sense a scientific notion or not. One can ask
whether health and its opposites can be given a neutral, rather than value-
laden description, or whether it follows by necessity that health is to be characterized as a 'good' bodily or mental state. Proponents of the former view are
often called naturalists, whereas proponents of the latter view are often called
normativists.

Different theorists have arrived at different conclusions with respect to this
issue. Boorse (1977, 1997) claims that there is a value-neutral definition of the
basic notion of disease. Wakefield (1992a) argues for the thesis that the notion
of disease has two parts, one of which is value-neutral, viz. the one that refers
to the natural function of organs. The other part of the concept, however,
refers to the value-laden notion of harm. Most other theorists, however, think
that the notion of health and its opposites is by necessity value-laden. Some
argue that these values are universal (Pellegrino and Thomasma, 1981), others
that the values determining the concepts of health and illness are connected to
the background cultures (Engelhardt, 1986). The physician/philosopher
Georges Canguilhem (1978), who wrote one of the most significant treatises
on human health and illness of the 20th century, though he drew almost exclusively upon medical data, came to the conclusion that health is an evaluative
concept in a strong sense. The healthy organism, says Canguilhem, is not an
organism whose functions are normal in a statistical sense. The healthy organism is one that is 'normative', i.e. one that is capable of adopting new norms
in life.

One can discern further differences in the contention that the notion of
health is value-laden. Some theorists (for instance Khushf, 2001) claim that the
notion is value-laden in the strong sense that its descriptive content can vary
over time. As a result of this, the only element common to an ancient and a
modern concept of health is that health is a 'good' state of a person's body or
mind. Others, like the ability theorists above, would claim that there is a
common descriptive content, viz. the fact that health has to do with a person's
abilities, but that one needs to make an evaluation in order to specify what
aspect or level of ability is required for health.

Health and Culture Relativism

If health is a value-laden concept then, as we have seen, some would argue that there are differences in the interpretation of health between cultures both historically and geographically. It is important to note that these differences can be more or less profound. The concepts of health can vary from culture to culture because there are fundamental differences in the basic philosophy of health and health care, as between Western medicine and traditional Chinese medicine or the traditional Indian Ayurveda medicine. Western medicine, which is to a great extent based on a naturalistic philosophy of man, arrives easily at a naturalistic understanding of health, whereas oriental schools with a holistic understanding of man in a religious context derive a notion of health which incorporates forces and developments that are partly supernatural.

The ways of and reasons for ascribing health to people may, however, vary even if there is a basic common theory of health and disease. Consider a particular physiological state, the state of lactase deficiency, which has the status of disease in a Western country but not in most North African countries. Lactase deficiency causes, in combination with ordinary consumption of milk, diarrhoea and abdominal pain. Thus, in Western countries where people ordinarily drink milk, lactase deficiency will typically lead to illness. Therefore this state ought to be included in a list of diseases in these countries. In North Africa, however, people rarely drink milk. Therefore lactase deficiency seldom leads to illness. Consequently it would be misleading to consider lactase deficiency a disease in this part of the world.

What makes the difference between the Western and the African cultures in this example is not different concepts of health and disease. It is a question of different lifestyles and different environments.

2

A Starting-point: Two Modern Streams of Philosophy of Human Health and Disease

Introduction

Although the variety of conceptions of human health, disease and illness is great, one can discern a sharper focus in the more systematic contemporary analyses of the concept of health and allied concepts. In fact, two main streams of theories of human health and disease have appeared in the arena. One of these is sometimes called the naturalistic stream, or the biostatistical one. What is typical of philosophers within this stream is that they claim that the concepts of health and disease and allied concepts – including illness, injury, impairment, defect, disability and handicap – are, or can be treated as, biological, or in certain cases psychological, concepts. 'Health' and 'disease' are biological concepts in the same sense as 'heart' and 'lung' and 'blood pressure' are biological concepts. In particular, there is, according to this position, nothing evaluative or subjective about the concepts of health and disease.

The other main stream in the philosophy of health involves a completely opposite position regarding these basic matters. According to these philosophers, who are often called normativists or holists, health and disease are intrinsically value-laden concepts. They cannot be totally defined in biological or psychological terms, if these terms are taken to be value-neutral. To say that somebody is healthy partly *means* that this person is in a good state of body or mind, the holist claims. And to say that somebody has contracted a disease is to say that this person has contracted something that is bad for him or her. (The relations between fact and value can be complex here. For instance, my own position with regard to the definition of health needs a more elaborate characterization. For an attempt to disentangle the relationship between fact and value with regard to health, *see* Nordenfelt, 2001, pp. 103–108.)

An example of a naturalistic definition is the following:

Health is the absence of disease. A disease is a type of internal state which is either an impairment of normal functional ability, i.e. a reduction of one or more functional abilities below typical efficiency [in relation to the biological goals of survival and reproduction], or a limitation on functional ability caused by environmental agents.

(Boorse, 1997, p. 8)

Examples of normative (holistic) characterizations:

A person is healthy when he is not suffering from evil and when he is unprevented from running his daily affairs.

(Galen, in Temkin, 1963, p. 637)

The patients who are ill are unable to do everyday things that people ordinarily just get on and do, moving their arms and legs, remembering ... things, finding their ways about familiar places and so on.

(Fulford, 1989, p. 149)

An example of a mixed naturalistic and normative definition:

A disorder exists when the failure of a person's internal mechanisms to perform their functions as designed by nature impinges harmfully on the person's wellbeing as defined by social values and meanings.

(Wakefield, 1992a, p. 373)

I will now turn to a further presentation of two theories of health and disease, each representing one of the streams.

Boorse's Biostatistical Theory of Health and Disease

On the biological side the articles by the American Christopher Boorse have been completely dominant in the arena. They have also been the target of most of the normativist counter-claims. In presenting Boorse's theory I shall use the most recent formulations made by Boorse himself in his long defensive article, published in 1997, called 'A rebuttal on health'.

The aim of Boorse's biostatistical theory of disease (BST) is to analyse the normal–pathological distinction. In order to capture the modern Western concept of disease Boorse proposes an explication of the ancient idea that the normal is the same as the natural in saying that health is conformity to species design. (Species design, however, is not interpreted here in the sense of evolutionary theory; see below.) In modern terms, Boorse says 'species design is the internal functional organization typical of species members, viz. the interlocking hierarchy of functional processes, at every level from organelle to cell to tissue to organ to gross behavior, by which organisms of a given species maintain and renew their life' (1997, p. 7). All conditions that are called pathological by ordinary medicine are disrupted part-functions at some level of this hierarchy, he says.

With this general description as a background Boorse presents the following definitions.

1. The *reference class* is a natural class of organisms of uniform functional design; specifically, an age group of a sex of a species, such as the human being.

2. A *normal function* of a part or process within members of the reference class is a statistically typical contribution by it to their individual survival and reproduction.

3. A *disease* is a type of internal state which is either an impairment of normal functional ability, i.e. a reduction of one or more functional abilities below typical efficiency, or a limitation on functional ability caused by environmental agents.

4. *Health* is the absence of disease. (Boorse, 1997, pp. 7–8)

The concept of normal functional ability is to be defined dispositionally – that is, as the readiness of an internal part to perform its normal functions on typical occasions with at least typical efficiency. Typical efficiency in its turn is an efficiency above some arbitrarily chosen minimum in its species distribution. The notion of a biological function that is crucial to Boorse's analysis is analysed in a value-free manner of the following kind. An organism or its part is directed to goal G when disposed, throughout a range of environmental variation, to modify its behaviour in the way required for G. And since physiology is the subfield on which somatic medicine relies, 'medical functional normality [is] presumably relative to the goals physiologists seem to assume, viz. individual survival and reproduction' (Boorse, 1997, p. 9).

What Boorse here presents is a pathologist's concept of disease. It is not the clinician's. Boorse does not analyse the state of affairs normally denoted by the term 'illness', which has a subjective component, the feeling of illness. Nor does he analyse any social or legal category, sometimes referred to as 'sickness', the social and legal role adopted by the person who is ill (Twaddle, 1993).

Boorse also claims that his characterization of disease is in line with what doctors generally mean by disease and what are accepted as diseases in current medical classifications. (In Nordenfelt, 2001, I have scrutinized and criticized this idea.) Finally, Boorse contends that his theory, although a descriptive theory, also contains refinements that go beyond existing medicine. 'My line on vagueness will be that the BST is never less precise than medical usage and fits all clear cases … Occasionally, the BST may be more precise than medicine, in which case it is a precising definition' (Boorse, 1997, p. 19).

An Action-theoretic Conception of Health and Disease

Some of the theories on the normativist (holistic) side also focus on goals, but they do so in a very different way. They do not refer to biological goals but to goals in the ordinary human sense, viz. goals of intentional actions. When we intend to do something or achieve something we automatically intend to attain a goal. Such a goal is not a goal of just a particular organ. It is a goal of the whole human being. Thus these theories are often called holistic theories.

It is significant that the holistic theories (or the HTH as I call them with a general term) consider the concept of health to be the primary one and that of disease a secondary one. Moreover, health has its basis on the level of the whole person. It is the person who is healthy, not the individual organs. Let me put this general idea of health in the old way once expressed by Galen, the famous Roman physician and philosopher from 200 AD: *Health is a state in which we neither suffer from evil nor are prevented from the functions of daily life* (quoted from Temkin, 1963, p. 637). My own refinement of this general idea is the following:

> A person A is completely healthy if, and only if, A is in a mental and bodily state, given accepted [or standard] circumstances, which is such that A has the second-order ability to realize all his or her vital goals, i.e. the states of affairs which are necessary and together sufficient for A's minimal happiness in the long run.
>
> (Nordenfelt, 1995, p. 93)

For a more complete presentation of this conception, see Part III of the present volume.

According to the HTH a person is to some extent ill when he or she does not fully possess such ability. A state of illness can have various causes within the person's body or mind. Such causes of ill-health as are common or typical are what we designate as diseases. Thus diseases, according to the HTH, are such bodily and mental states of affairs as tend to lead to their bearer's ill-health.

Two kinds of phenomena have a central place in traditional holistic accounts of health and illness. First a certain kind of feeling, of ease or well-being in the case of health, and of pain or suffering in the case of illness; second the phenomenon of ability or disability, the former an indication of health, the latter of illness. These two kinds of phenomena are in many ways interconnected. There is first an empirical, causal connection. A feeling of ease or well-being contributes causally to the ability of its bearer. A feeling of pain or suffering may directly cause some degree of disability. Conversely, a subject's perception of his or her ability or disability greatly influences his or her emotional state.

In my own analysis (Nordenfelt, 1987/1995) I make an assumption of a strong connection between suffering and disability, where suffering is taken to be a highly general concept covering both physical pain and mental distress. A person cannot experience great suffering without evincing some degree of disability. But the converse relation does not always hold: a person may have a disability, and even be disabled in several respects, without suffering. There are, for instance, paradigm cases of ill-health where suffering is absent. One obvious case is that of coma, when a person does not feel anything at all. Another concerns certain mental disabilities and illnesses. In general, when patients cannot reflect properly on their own situation, then their disabilities need not have suffering as a consequence. In short, therefore, wherever there is great suffering there is disability, but the converse is not true.

These observations indicate that the concept of disability should have a more central place in the *defining* characterization of ill-health than the corresponding concept of suffering. If one of these notions is essential to the concept of ill-health it must be disability. This conclusion does not deny the extreme importance of pain and suffering – as experiences and not just as causes of disability – in most instances of ill-health.

Protagonists of holistic theories of health often, unlike Boorse, attempt not only to capture the doctor's use of terms such as 'health' and 'disease' but also the way lay-persons use these terms. Moreover, they are interested in explaining the web of concepts that surrounds the ordinary concepts of health and disease. Another purpose that is salient in the work of Fulford (1989) and Nordenfelt (1987/1995) is to contribute to the amendment of the prevalent medical conceptual network. My own view, for instance, is that the current conceptual network is deficient. For instance, notable ambiguities pertain to the use of 'illness', where that term is sometimes used synonymously with 'disease' and sometimes not. Such ambiguities do not contribute to the good cause of medicine.

Let me now summarize the two theory types in the following formulae:

1. The BST. The individual A is completely healthy if, and only if, all organs of A function normally, i.e. make their species-typical contribution to the survival of the individual and the species, given a statistically normal environment. A disease is a subnormal functioning of a bodily or mental part of the human being.
2. The HTH. A is completely healthy if, and only if, A is in a bodily and mental state which is such that A is able to achieve all his or her vital goals, given standard circumstances. A disease is a bodily or mental process that tends to reduce the health (as holistically understood) of the human being.

In Part III of this book I will return to a deeper analysis of and comparison between the BST and the HTH.

A Mixed Theory of Health: Jerome Wakefield

A naturalistic line of thought rival to the BST entered the arena during the 1990s. According to this thinking the notions of health and illness are to be put within the context of evolutionary theory. The notion of biological dysfunction is crucial here. According to this theory, a condition must be a dysfunction or be caused by a dysfunction in order to qualify as an illness. The major protagonist of this evolutionary line of thought, Jerome C. Wakefield, has promoted his ideas mainly within the current debate on the nature of mental disorders and his theory has therefore not become so well-known in general philosophy of medicine. It is however clear that his theory, if valid, could also be applicable to the characterization of somatic health and illness.

As we have seen, the notion of function is central also to Boorse. However, Boorse and Wakefield develop their ideas differently. Boorse abstains from referring to biological evolution in making his characterization of health. Boorse's theoretical characterization of the natural is exclusively related to the present context: the healthy organ is the one that makes a statistically typical contribution to the two main biological goals, viz. survival of the individual and survival of the species. Jerome Wakefield, on the other hand, underlines the historical evolutionary context in explicating the notion of a natural function, which is the crucial notion for his characterization of medical notions such as health and illness.

The starting-point for Wakefield's analysis is the state of psychiatric theory around 1990. At that time there was much controversy concerning the status of the *Diagnostic and Statistical Manual of Mental Disorders* (DSM) and the criteria that the constructors of this manual used in selecting conditions suitable for registration in it. Since the third version of DSM (III) from 1980 there has been a definition of mental disorder in the manual. It runs as follows:

> Each of the mental disorders is conceptualized as a clinically significant
> behavioural or psychological syndrome or pattern that occurs in an individual and
> that is associated with present distress (e.g. a painful symptom) or disability (i.e.
> impairment in one or more important areas of functioning) or with a significantly
> increased risk of suffering death, pain, disability or an important loss of freedom.
> (DSM IV, 1994, p. xxi)

There was, however, at the time a lot of complaint that this definition did not play the role that it ought to play in distinguishing pathological from non-pathological disorders. This, some critics claimed (Kirk and Kutchins, 1997), was partly due to the fact that the definition itself needs interpretation and amendment. This is the task that was undertaken by Wakefield in a number of seminal articles during the 1990s.

Wakefield (1992b) starts with an improved and simplified version of the DSM definition: a mental disorder is a mental condition that: (i) causes significant distress or disability; (ii) is not merely an expectable response to a particular event; and (iii) is a manifestation of a mental dysfunction. Wakefield thinks that (i) and (iii) are the crucial and valuable elements in a definition of mental disorder; (ii) is, he thinks, either redundant or misleading. There are, he thinks (and, I think, correctly), instances of mental disorder that are expectable. If a person is exposed to a sufficient amount of stress and damage, then the expected outcome certainly is disorder, whether physical or mental. Thus, the expected outcome cannot be a disqualifier. Both harm and dysfunction, however, are essential elements in Wakefield's characterization of mental disorder. This means that Wakefield's position in the theory of health is complex. It has both naturalistic and normative elements. A mental disorder is *both* a defect from the point of view of natural order *and* a harmful state of affairs. In short, a mental disorder is a harmful mental dysfunction. More expanded: 'A disorder exists when the failure of a person's internal mechanisms to perform their functions as designed by nature impinges harmfully on the person's well-being as defined by social values and meanings' (Wakefield, 1992a, p. 373). But the dysfunction requirement 'is necessary to distinguish disorders from many other types of negative conditions that are part of normal functioning, such as ignorance, grief and normal reactions to stressful environments' (Wakefield, 1992b, p. 233).

Wakefield says rather little about the harm aspect of the mental disorder.

It is, as he says, to be translated into a list of specific and recognizable kinds of harm, particularly stress and disability. This he has in common with almost all normativists in the theory of health. However, in one respect he takes a stand in favour of a social interpretation: 'only dysfunctions that are socially disvalued are disorders' (Wakefield, 1992a, p. 384).

The Notions of Dysfunction and Natural Function

Wakefield's preliminary specification of the notion of dysfunction is the following:

> A dysfunction exists when a person's internal mechanisms are not able to function in the range of environments for which they were designed. Thus, one can construct a test for dysfunction by specifying an environment in which the function is designed to manifest itself; if the function fails to be manifested in that environment, there is likely a dysfunction.
>
> (Wakefield, 1992b, p. 243)

A dysfunction, Wakefield says, should be related to the natural function of an organ in an organism. Every organ has, as we intuitively think, a natural function. The most celebrated example is the function of the heart, which is taken to be to circulate the blood in the body of the animal in question. Sometimes an organ has more than one natural function; the liver, for instance, both realizes glycogen synthesis in the body and purifies the blood from various poisons. But how should this notion of natural function be analysed? There are various slightly different attempts to be found in the literature. I shall here focus on Wakefield's formulations.

A careful analysis of the concept of natural function, says Wakefield, leads to the conclusion that the most viable approach to it is based on the notion of *evolutionary design*.

> Certain mechanisms are naturally selected because of the beneficial effects that the mechanisms have on the organism's fitness, and those beneficial effects are the natural functions of those mechanisms. The natural function, then, is not just any benefit or effect provided by a mechanism but a benefit or effect that explains through evolutionary theory why the mechanism exists or why it has the form that it does.
>
> (Wakefield, 1992b, p. 236)

A crucial question here is the following: is there a coherent and reasonable concept of natural function? And if there is, does it have anything to do with a viable notion of illness or disorder or, for that matter, with the notions of welfare and illfare? First a few ontological points. There is sometimes confusion between the notions of function, functional ability and the exercise of a functional ability. Not infrequently these terms are used interchangeably. Let me make the following remarks concerning my own use.

An organ may have a function, and to say this is to make a statement on the *species* level. The function of the human heart is to reach a certain state of affairs or to keep a process going. The function then relates to a result of the working of an organ. The organ of an *individual* human being may have an ability to realize this function, or it may not. In the latter case it is dysfunctioning. And an individual organ may or may not at a particular point exercise its functional ability. Some organs exercise their functional ability all the time, as the heart does. Some organs, such as the stomach or the kidney, exercise it only at certain times.

Let me now go a little into the basic idea of a natural function. Start with the heart. The heart pumps blood. The result of the pumping, the distribution of the blood, is the natural function of the heart in a human. What does this mean? The heart has this function, says Wakefield, because the human heart – like of course all hearts of animals – was *designed* in this way and for this purpose.

But design is metaphorical talk, admits Wakefield. The terms 'species design' and 'selection' are metaphors, as are several other terms in this area of discourse. They are chosen in analogy with the human case of artefacts where a human being intentionally designs and selects. No such thing, or indeed any analogous assumption related to the intentions of God, is assumed in the theory of evolution. This is underlined by Wakefield and others who put forward the evolutionary argument.

The idea of design must have a non-intentional interpretation in the field of biology. And this non-intentional interpretation is a causal interpretation. There is a particular causal history of the existence of the heart in a human being. And this causal history crucially involves the distribution of the blood. At some stage in the early history of mankind there were individuals quite similar to us. They had hearts and lungs that worked in a way that was similar to the way hearts and lungs work with us. These hearts and lungs contributed to the survival of the individuals at that time. And since humans had a number of properties that together with the work of hearts and lungs made them fit to survive as a species, given the environments existing up to our time, we still have humans on the Earth. In all human individuals today there is a heart that pumps blood and thereby distributes it around the body. Since the similar work of hearts long ago contributed to the survival of humans, then we can designate the distribution of the blood to be a natural function of the human heart.

So much for the introduction of the idea of a natural function. The specification of a dysfunction is then easy. An organ or a mental faculty has a dysfunction, or is dysfunctioning, if, and only if, this organ or faculty does not (completely) fulfil its natural function as specified above. The dysfunction is then a property of an individual within a species. It is a *contradictio in adiecto* to talk about a species that is dysfunctioning. This observation has some bearing on the discussion about the relation between dysfunction and disorder.

3

The Place of Evolutionary Theory in the Philosophy of Health and Welfare

The Problem of Identifying a Natural Function

The idea of a natural function is central to several theories of health and welfare both in the human and the animal arena. (See, for instance, McGlone, 1993 and Barnard and Hurst, 1996 below.) Frequently this idea is interpreted within the framework of evolutionary theory as indicated in the case of Wakefield. It is doubtful, however, whether evolutionary theory can be directly transposed to the theory of health and welfare. I will raise a few issues here.

First, not all so-called functions are as easily delineated as the distribution of blood in the case of the heart. This function is salient in the sense that the distribution of the blood is necessary for the survival of the organism at all moments of its life. If the heart stops pumping, the organism very soon dies. The distribution of the blood is a strictly necessary condition of the individual's survival today. One can then very plausibly presume that this was a necessary condition for the survival of humans and previous anthropoids long ago. And if something is a necessary causal factor in respect of the existence of every individual belonging to a particular species, it is a partial causal factor in respect of the continued existence of this species today.

But what shall we say about parts of the body where we do not at all have initial intuitions about a natural function? This actualizes an important problem concerning the idea of historical evolution. This idea focuses on what functions were crucial to humans *in the past*, what made us survive in the past, given *the standard circumstances in the past,* or even given some very special circumstances in the past. Consider the following. It may have been the case that the human species was dying and almost extinct 100,000 years ago. Humans (or our anthropoid ancestors) were, according to this assumption, really not at all fit for the climate and for the general ecological situation at that time. The so-called natural functions of those humans were not viable in the context.

However, because of a climate change a pocket population was rescued and these humans could start building up a viable population again.

Thus, plausible speculations concerning the benefits of certain functions and faculties for the survival of humans *today* cannot be immediately applied for establishing something's status as a natural function. One can certainly see that *today* it is important for a human to grasp things, or to reason about things and to memorize things, to take two mental examples. The ability of memorizing things is of particular importance today, when we, for cultural reasons, are forced to remember a large quantity of number combinations. But this ability need not have been crucial in olden times.

But the fitness of a function today is *not* the criterion of its being a natural function according to Wakefield's definition. One can at most get a clue for identifying a natural function by considering its survival value today. It is the fitness through history that is the criterion; in fact the function in question may be of limited or no use today and still be a natural function. There are probably some physical capacities that are more or less superfluous today but were quite important long ago in the old human societies of hunters and fishermen. This, then, marks the important interesting difference between Wakefield's notion of natural function and Boorse's notion of normal function. The latter notion is completely focused on the beneficial strength of the function today. Boorse says: the normal (thus: healthy) function is the one that makes a typical contribution to the individual's survival, or the survival of the species, today.

The specification of what are the natural functions of the human organism (as well as of all other organisms) is then not just a difficult affair – it is strictly speaking an impossible affair. With the exception of the most salient functions, such as the main functions of the heart and the lungs, we are, and must remain, too ignorant to tell whether a particular present disposition of an organ or a faculty should be labelled a natural function or not. This is not a promising state of affairs for the usefulness of the notion of natural function (given the evolutionary interpretation) as a criterion of health.

Similarly, one can wonder how the historical evolutionary perspective can help us determine what the natural or proper *degree* of an alleged human function is. How much blood should the heart pump? How well should the hand grasp? How much should one be able to memorize? How much sorrow should one feel when a relative dies? What resources, other than statistical ones, do we have to determine what is a proper degree?

Since the environment, in terms of climate and other natural conditions, has changed so considerably over time, the demands on people have changed greatly. At one time, at least humans in the north had to withstand cold and to be able to survive on very few nutrients. At some other time humans may have been particularly forced to compete with other animals in predation. At some further stage, humans may have lived quite sheltered lives and with an abundance of food. So the requirements have varied over time and place. Thus assume that at some time t function F to degree 10 was necessary for the survival of a population, at time t1 F was necessary to degree 8, and now it may be necessary to degree 4 for the survival of the present population. What is then the 'proper' natural function as quantitatively specified? Which historical stage shall we choose for this decision? We can hardly use the present level as that goes against the importance of the evolutionary perspective: the function should be something that has been historically selective. But what segment of history shall we choose? The choice seems to be arbitrary. Let me on this point quote one of the most prominent evolutionary biologists, Ernst Mayr:

> Organisms are doomed to extinction *unless they change continuously* in order to keep step with the constantly changing physical and biotic environment. Such changes are ubiquitous, since climates change, competitors invade the area, predators become extinct, food sources fluctuate; indeed hardly any component of the environment remains constant.
>
> (Mayr, 1982, pp. 483–484)

There is a plausible reply to this observation. According to this idea it is not really the case that the 'real' functions change or that the proper degree of the real functions changes over time. The changes occur instead on a more superficial level. Take the human heart as an example. The heart pumps blood. It is elementary that the pulse varies a lot with the environment and with the activity of the subject. One cannot say, therefore, that a particular frequency of the beating of the heart is the 'natural' frequency. The frequency in a way compensates for changes in the environment, in order to attain the crucial result of distributing blood and thereby oxygen to all other vital organs and tissues of the body in the required manner. Thus, the pulse is not in itself a 'real' function according to this reasoning. The real function is the distribution of oxygen to other organs and tissues. And this may be performed equally well in different kinds of situations.

The problem with this argument is that we can observe just as many possible differences on the higher level, viz. regarding the distribution of oxygen. The environment does not just require different pulses but also different levels of oxygen for different purposes and different contexts. Hence the blood distribution also becomes disqualified as a 'real' function. We must look for the real function on an even higher level of the organization of the human being. But also on that level we can find differences and no 'real' functions. Thus, the seeking for the real function will continue until we end up on the completely abstract level, where we say that the hearts in the different environments make equivalent causal contributions to the survival of the individuals in question. This may be true, but we have then completely capitulated in our endeavour to find functions on the concrete level of the organs where we wanted to find them for the purpose of detecting particular dysfunctions.

On the Scope of the Notion of Natural Function

The problem of the free riders

Let me now also note another complicating factor with regard to the notion of a natural function. We understand that the fulfilment of a particular function among the individuals cannot be the sole cause of the survival of a species (or a certain population). A complex web of factors makes it possible for a population to survive given a certain environment. But in addition to such factors as undoubtedly contribute to the survival of the species there may be properties of a body or mind which also survive, partly because they are causally connected to the crucial factors, but which do not contribute positively at all to survival. These factors are in a way *free riders* in the evolution of the species. (For a similar observation, see Stephen Jay Gould, 2000. He there talks about important evolutionary features that do not arise as adaptations. He says: 'Since organisms are complex and highly integrated entities, any adaptive change must automatically "throw off" a series of structural byproducts … Such byproducts may later be co-opted for useful purposes, but they didn't arise as adaptations', p. 95. See also Gould and Vrba, 1998.)

We can confidently assume that we have some evolutionary free riders existing in our bodies and minds today. We normally assume that this is the case with the human appendix, for instance. The free-riding organ typically does not contribute to the survival of the organism. It could be completely neutral in this respect. But we can also imagine cases where members of a species have survived *in spite of* a certain trait that they have had and still have. Other traits and properties can have compensated for a relative weakness of the trait in question.

The interesting question for the natural function theorist then is: what shall we say about the free riders, the non-contributing organs? Do they not have any natural functions? Can we not find any dysfunctions in such organs? This seems to follow. Consider again the characterization made by Wakefield: 'Certain mechanisms are naturally selected because of the beneficial effects that the mechanisms have on the organism's fitness, and those beneficial effects are the natural functions of those mechanisms'. The working of the free-riding organ has per definition not been selected because of the benefits of its function for the survival of the organism. Thus this working cannot be an instantiation of the exercise of a natural function of the organism in question.

What consequences do these answers have for a theory of health or, as we shall see, a theory of welfare? If some organs or faculties do not have any natural functions, does this mean that they cannot have any diseases or disorders? Let me here assume that the appendix in the human body does not have a natural function. There are continuous processes, however, in every living individual's appendix. But, per definition, there cannot be any dysfunction in the appendix. Thus the appendix cannot have any disease. But what about an infection in the appendix and what about pain in general that comes from the appendix? Do these not constitute blatant instances of diseases?

There is one good, but I do not think sufficient, answer to this argument. It runs as follows. Most diseases in the free-riding organs have negative consequences for other organs, including evolutionarily selected organs, and for the functional ability of these organs. An infection in the appendix, for instance, may cause general disturbance in the functional ability of the organism; it may even be lethal. Thus, it is easy to see how the infection in the appendix can qualify as a disease, even under the dysfunction interpretation. The disease must then, however, strictly speaking be located outside the appendix.

I admit that this answer covers several cases of diseases that we normally locate in free-riding organs. But I contest the assertion that it covers all cases. The paradigm contrary case is the local disease that causes pain without otherwise disturbing the major parts of the organism. A free-riding organ may be damaged through an accident, causing considerable pain and preventing the person from going to work. Here, by assumption, there is no general infection in the body, and there is no other general physiological change.

A tentative counter-argument is the following: also in this case there is a general disturbance of the organism, in that the pain and disability itself constitutes a general disturbance. This reply is, however, a capitulation. If the answer is that pain and disability can qualify as dysfunctions, then it becomes impossible to uphold Wakefield's distinction between the dysfunction component and the harm component of a disease. Pain and disability are the paradigm features of harm in the health context. The harm component must be distinguished from the internal dysfunction.

I conclude therefore that the argument from free riders is a forceful argument against the idea that dysfunction (in the evolutionary sense) is a necessary condition of disease or disorder.

The problem of the dying species

Consider also the following problem. Some zoologists find a new species on the Galapagos Islands. This is a species, they find out, some members of which have managed to survive in this pocket of the Earth, given the very peculiar circumstances of the Galapagos Islands. However, now civilization and its consequences have caught up with them. The species is clearly dying. The population is in the long run not capable of protecting itself against pollution and new predators. All experts expect the remaining population to die within a decade or so. However, so far this population has survived, given a combination of properties of its members and the peculiar relatively stable environment which has existed up till now. Thus, on the causal interpretation of natural function the members of the dying species indeed have natural functions, viz. the ones that have in the past caused them to survive. But these very functions, and the attempts to fulfil them by individual members, are now in effect killing the remaining individuals. They contribute to the extinction of this species. Shall we then conclude that the features which are killing these individuals represent the natural order and thereby the healthy working of these organisms?

Also here one can expect a counter-argument from the natural function theorist. A set of natural functions (or of 'healthy' functions) cannot protect an individual or indeed a population from extinction if there is a natural catastrophe. The circumstances can be so harsh that also (intuitively speaking) healthy individuals have to succumb. One can therefore not use a catastrophe as an argument against an evolutionary theory of health.

I grant this argument for the case of the real catastrophe, viz. a sudden event, for example the falling of a meteor to Earth, which destroys and kills members of a population. My example of the dying species on the Galapagos

Islands, however, does not represent such a catastrophe. Instead it is intended to refer to a rather slow process, say over some hundred years, during which a population has not been able (or had the time) to adapt itself. This kind of process is not a sudden catastrophe like the falling meteor, it is instead a gradual change of the standard circumstances of the Earth, viz. such circumstances as every living being will have to cope with in the foreseeable future. These are the circumstances, to return to the human case, which human beings take as their norm when they subjectively evaluate their own health.

The upshot of the argument from the dying species is that individuals of a dying population can remain with their natural functional abilities intact. And as long as these are intact we cannot, according to the evolutionary theory, ascribe illness or disorder to the members of this population.

4

Two Classic Theories of Human Welfare or Happiness

I will now turn to some theories of human welfare or quality of life. First, I will introduce two classic and quite different ideas, Aristotle's theory of *eudaimonia*, and Bentham's utilitarian view of welfare. I will thereafter consider some modern attempts to characterize and operationalize the notion of welfare or quality of life, mainly in the context of health care.

Some Main Points in Aristotle's Theory of *Eudaimonia*

The ultimate good in life was called *eudaimonia* by Aristotle. Literally this means 'being blessed with a good daimon'. The latter in its turn means 'a divine guard' (*The Eudemian Ethics*; for a modern edition, see Aristotle, 1982, p. 48). This literal translation can give one the impression that Aristotle by *eudaimonia* means that the person who is favoured and guarded by the gods and who in general lives in good circumstances, lives the best life. Such an interpretation, however, would give a misleading picture of Aristotle's theory.

It is difficult to translate *eudaimonia* into the Western languages of today. The traditional translation in English, however, is 'happiness'. It is then immediately important to realize that Aristotle by happiness means something quite different from (although partly connected with) what we today normally mean by happiness. A first crucial difference between the *eudaimonia* of Aristotle and the happiness of today is that *eudaimonia* is not a state of a person but an activity. Aristotle does not regard happiness as a state that one is in possession of. Nor is happiness an experience. Instead happiness consists in a certain type of active life. On the other hand, a crucial similarity also exists between Aristotle's *eudaimonia* and the kind of happiness that, for instance, the utilitarian ethics of our time speaks about. Both concepts have an equally central place in ethics. *Eudaimonia* and (utilitarian) happiness are the things to be achieved, both for oneself and for others, according to the corresponding ethical theories. *Eudaimonia* and happiness are both the goals of human activity. We must then, however, bear in mind that *eudaimonia* is in itself an activity.

But what is the activity that Aristotle considers to be the meaning and goal of life? What should one pursue in order to live the best conceivable life? To answer these questions one has to presuppose certain other crucial parts of Aristotle's theoretical philosophy. Aristotle believed, and tried to persuade his readers, that his theory of welfare follows logically from his basic metaphysics. The commentators of our time are perhaps not so convinced that this is so. One major difficulty is that Aristotle's ideas in this area are expressed in slightly different ways in his two major ethical works: the *Nicomachean Ethics* and the *Eudemian Ethics*. In spite of this let me here try to reconstruct the main steps in Aristotle's reasoning. (For the sake of simplicity I shall exclusively refer to the most famous of the two works, viz. the *Nicomachean Ethics* (NE); for a modern edition, see Aristotle, 1934.)

An important ingredient in Aristotle's theoretical philosophy is that all entities in the world have a *function*. When we deal with artefacts this idea is rather simple and understandable. A knife has the function to cut things; an axe has the function to split wood; a sewing machine the function to sew, etc. But Aristotle's idea is quite general. All entities, be they living or dead, have some function. With regard to the biological world we might express this thought by talking about the animals and plants in ecological equilibrium. Plants and animals are preconditions of each other. Plants constitute by their photosynthesis and their production of oxygen a general precondition of animal life. Small animals, like insects, constitute food in particular for birds. These in their turn are food for predators, and so on.

The function of an object, Aristotle says, is a part of its essence. The ability to split wood belongs to the essence of the axe. According to a modern way of speaking we might say that it belongs to the definition of the concept of an axe that it is an object by means of which one can split wood. Could the object not split wood, then it would not be an axe. In the same way we could reason concerning other objects, including animals and human beings (NE, Book I, vii, 10–16).

The functions are hereby distinguishing features of objects. The function of the axe is distinct from the function of the stone. The function of the anemone is at least partly distinct from the function of the orchid; similarly with the function of the buzzard, the chimpanzee and the human being.

Another crucial thought behind Aristotle's view of the good life is that the purpose of every object is to exercise its function. It is therefore essential that the living being realizes itself, thereby exercising its function, and that the inanimate object is used for its purpose, thereby exercising its function. Here we have a crucial premise in Aristotle's reasoning: it is better for an object to exercise its function than to be inactive. Therefore it is better for the human being to actively exercise his or her function than to live in a passive state. Therefore, *eudaimonia*, the best thing in life, must be an activity and not a static state.

The purpose of the existence of a human being therefore must be to exercise the function typical of human beings. It is then easy to add that the good life is to exercise the function well. The best conceivable life for a human then is to exercise his or her function in the best conceivable way.

But what is the function in the case of the human being? Are we only dealing with one function? Here we enter controversial territory, but certain things seem to be clear. That or those functions that constitute the essence of the human being must be distinct from the functions that apply to other living beings. Otherwise we would not be able to separate human beings from these other living beings. The function of the human being cannot, for instance, be the intake of nutrients. This activity we have in common with both plants and other animals. Nor can the human function be just to breathe, move about or be sexually active. This we have in common with all other animals. We must find something that is specific for *Homo sapiens*, the 'sage' man (NE, Book I, xiii, 5–20).

What Aristotle focuses on is that human beings, in contradistinction to the animals, can act according to rational principles, which he takes to be equivalent to acting according to norms of virtue. To live virtuously is the same as exercising the human function. A virtuous human being, in the sense that he or she is also active and continuously practising his or her virtue, thus lives a life in *eudaimonia* and is happy.

This is not Aristotle's complete answer, however: he also thinks that there are different degrees of virtuous life and that there can be different degrees of *eudaimonia*. This grading of *eudaimonia* is not just dependent on how often or to what extent one performs virtuous actions. No, the degree of *eudaimonia* also depends on the nature of the virtues exercised.

Aristotle proposes a hierarchy of virtues in the following sense: such virtues as are predominantly of a spiritual character, or such as are mainly performed by the soul, have a higher position in the hierarchy than such as are mainly of a bodily kind. To perform well with one's body is good, but it is even better to perform well with one's soul, i.e. with that part of the person that distinguishes him or her from the animals. In fact, this is the best conceivable kind of activity. The complete *eudaimonia* therefore consists in contemplation, in particular the contemplation of abstract ideas (NE, Book X, vii).

To this we can add a point concerning time. In order for real *eudaimonia* to occur, such contemplation should apply to a complete lifetime. Strictly speaking, it is only the completed life that can be referred to as happy in the Aristotelian sense (NE, Book I, vii, 16). So much for the general framework of Aristotle's philosophy.

One may think that Aristotle's theory of happiness is almost absurdly unrealistic. One cannot conceive of a human being who can contemplate supernatural truths throughout life. Is it then reasonable to propose such a theory? Aristotle's answer to such an objection would probably include reminding us that continuous contemplation is the highest (utopian) degree of *eudaimonia*. He does not mean that all other lives completely lack *eudaimonia*. On the contrary, many people can, according to Aristotle, reach a high degree of happiness without reaching the ideal. In fact, all philosophies of happiness would need a similar thesis. No human being can in practice be completely happy. But this does not preclude us from giving a theoretical characterization of the notion of complete happiness (NE, Book X, vii, 9).

Aristotle also reminds us that there exists not just an abstract description of human *eudaimonia* in general. There are also concrete descriptions of the *eudaimonia* of every specific individual. Different individuals have different degrees of potential *eudaimonia*. Some people have small intellectual capabilities and therefore have little possibility of exercising intellectual virtues. Their competence to exercise other virtues can, on the other hand, be much greater. Conversely, one can say that an intellectual who is bedridden has little competence to live such a virtuous life as presupposes bodily activity. On the other hand, his or her opportunity for living a contemplative life is the greater. These two different extremes of human beings have through their constitution completely different potentials for *eudaimonia*. But in both cases they would probably be rather far from the highest conceivable *eudaimonia*.

In all realistic discussion, whether it deals with morals or with the prudent planning of a human life, these basic presuppositions have to be considered. For each individual one has to try to characterize that degree and kind of *eudaimonia* that is at all attainable for him or her. The crucial thing, then, according to Aristotle, is not that one shall exploit as many of one's potential activities as possible and thereby act in a bad or non-virtuous way. The crucial thing is that such activities as are actually performed are performed well. It is better to be a skilful carpenter than a mediocre author. It is better to be a skilful soldier than a scarcely competent prime minister. These are some of the conclusions to be drawn from the theory of Aristotle.

But it is not only the constitution of a person that confines his or her ability to reach *eudaimonia*. Also external circumstances do this to a high degree. It is clear that the person who lives in impoverished and appalling circumstances does not have the same opportunity to live a virtuous life as the rich person who lives in a peaceable and creative society. It is not possible to be generous if there is nothing to give; it is not possible to be a contemplative person if one is a parent of a great number of hungry children who continuously crave food. Thus it is not only happiness in our modern sense that is confined by negative circumstances. This holds equally well for Aristotle's *eudaimonia* (NE, Book I, viii, 15–17).

Aristotle was quite clear – in fact much clearer than many contemporary debaters – about the difference between the conditions for *eudaimonia* and *eudaimonia* itself. Many conditions, both external and internal, must be fulfilled in order for a person to reach a high degree of *eudaimonia*. Among these conditions are a certain minimal degree of health, a minimum of physical protection and a minimal economic platform. Perhaps one should add a minimum of social life. But these necessary conditions are not together sufficient for *eudaimonia*. In order for a person to have *eudaimonia* he or she must act; the person must be active, and be active according to a rational and virtuous principle.

Thus I can repeat the thesis that I proposed at the beginning of this section. To have external and internal welfare is not the same as having *eudaimonia.* The happy person must also act on the basis of this welfare. This fact indicates the difference between *eudaimonia* and the modern Western concepts of happiness. If happiness in the modern sense is a kind of experience or disposition towards experience, then *eudaimonia* is something completely different. Our modern version of happiness is a mental state and not an activity.

In some of our modern concepts of happiness sensual pleasure is a species of happiness. That is the case with Jeremy Bentham's concept (see the next section). This is far from Aristotle's view. He has a specific argument for excluding pleasure from *eudaimonia.* To feel pleasure is not specifically human. Many other animals have the capacity to do so. Thus, pleasure cannot be a part or species of *eudaimonia*, which is the function of man. Nor does it follow from Aristotle's concept of *eudaimonia* that a person who is quite happy must have a high degree of *eudaimonia.* Or the converse: from a high degree of *eudaimonia* does not follow logically a high degree of experiential happiness. On the other hand, Aristotle seems to have meant that the virtuous and active person has normally as a matter of fact a high degree of experiential happiness. If one happens to be interested in the latter kind of happiness – which we should not, according to Aristotle – it is a good recipe to opt for *eudaimonia* in order to attain it. The virtuous person is normally quite a happy person in our modern sense. The happiest person is the contemplative philosopher, who is continuously in contact with the supernatural facts.

Jeremy Bentham and the Utilitarians

The activity-oriented theory of the good life that was advocated by Aristotle has not much influenced the welfare philosophers of our time. I have in mind primarily the utilitarians who have been the leading figures in the Anglo-Saxon countries and Scandinavia.

The modern forefather of utilitarianism is Jeremy Bentham (1748–1832). He presented his main ideas on moral and legal matters in the great work *An Introduction to the Principles of Morals and Legislation* from 1789 (a recent edition 1982). I will here present some of his main thoughts.

According to Bentham our life is governed by two main principles, the principle of pleasure and the principle of pain. We all have a natural tendency to aim for pleasure and avoid pain. In fact, says Bentham, all our voluntary actions are ultimately motivated by our desire to seek pleasure and avoid pain. This does not entail that such a desire has to be the immediate motive behind all actions. What Bentham intends is that ultimately, in that chain of causes which results in an action, there must either be a desire for pleasure or a desire to avoid pain. By stating this Bentham shows that he is an extreme representative of the school of *psychological hedonism.*

So far, however, I have only talked in psychological terms. I have said that there is a human tendency to behave in certain ways. But what has this to do with morals or with the ideal human life? Does the biological natural tendency have to be the morally correct one? *Should* we aim for pleasure and avoid pain?

Yes, in a way this is what Bentham and the other utilitarians mean. We ought to follow our psychological inclination and let it constitute the foundation of our morals. Bentham claims that it is both unnecessary and impossible to prove that such an idea is true. The thesis that we ought to aim for pleasure and avoid pain is a thesis that resembles the axioms in mathematics. It is used to prove other things but cannot itself be proved. (Bentham's famous pupil John Stuart Mill tried to prove the utilitarian principle. Most commentators agree, however, that he failed in this attempt. For a scholarly commentary *see* Harrison, 1983, p. 168.)

But the classic utilitarian principle is not merely that we should aim at pleasure or happiness for ourselves. It is more general: we should aim at the greatest happiness of the greatest number of people (Bentham, 1776, *A Fragment of Government*; a recent edition 1977). This is a further step, and how could that be theoretically justified? When Bentham says that we all have a natural tendency to seek pleasure and avoid pain, he must be referring to the subject's own pleasure and pain, i.e. some kind of egoism. But the ethical doctrine advocated by Bentham and later, most vigorously, by John Stuart Mill talks about how we should act towards the whole of mankind. The self is not excluded but it does not have a privileged position.

According to authoritative interpretations (for instance in Bentham, 1982, pp. xlvii–xlviii) Bentham reasons in the following way: a rational person realizes that his or her own long-term interests will probably benefit most if he or she subscribes to a moral code such as the utilitarian. Admittedly, this will now and again entail that the person's short-term interests have to give way to the interests of others. But succumbing to these restrictions is a precondition for motivating other people to embrace the same moral code. And a universal support for this code is in its turn a precondition for a harmonious society, which in its turn forms the basis for the subject's satisfaction. According to Bentham, a rational person will accept a utilitarian ethics by performing this reasoning that has psychological hedonism as its starting-point.

So far a few words about utilitarianism as an ethical theory. I will now, however, focus on Bentham's views on the nature of pleasure or happiness and of the various kinds of pleasure or happiness that he acknowledges. I will also briefly comment on his ideas on measuring pleasure or happiness. These are in a way the classic forerunners of the measurements of quality of life today.

First some remarks about terminology. Bentham does not make a clear distinction between pleasure and happiness. On many occasions he talks about 'pleasure or happiness' as if the two terms were almost synonymous. This means that pleasure with Bentham is something extremely general, referring to all kinds of positive sensations, moods and emotions. This general use of the term is not common today. (Moreover I will myself make a sharp distinction between pleasure and happiness in my own proposal for a theory of welfare and quality of life; see Chapter 22.) I will in this section, however, follow Bentham's mode of speech and talk about pleasure in a general sense.

I will for similar reasons keep the term 'pain' for the general negative counterpart, although 'pain' today almost exclusively refers to a bodily sensation.

The different kinds of pleasure and pain in Bentham's theory

Bentham specifies a number of dimensions according to which a measurement
should be made. An experience of pleasure or pain should be characterized
with regard to its position along the following scales:

1. Intensity.
2. Duration.
3. Certainty or uncertainty.
4. Propinquity or remoteness.
5. Fecundity (i.e. ability to generate new sensations of the same kind).
6. Purity.
7. Extension.

Unfortunately Bentham offers rather few comments on these dimensions. It is
sometimes, therefore, unclear how to interpret them. The lack of clarity concerns primarily the dimensions certainty and propinquity. Does the certainty of
an experience of pleasure refer to whether the experience is precisely an experience of pleasure? Or does the certainty have to do with the stability of the
experience, i.e. it will last for some time? Should we by propinquity mean
whether the experience is close to one's consciousness, i.e. is more or less
present in one's consciousness? Or does Bentham refer to the possibility that
it is easy to recall the feeling to one's consciousness?

The dimensions of purity and extension, however, are given a clear interpretation by Bentham. An experience of pleasure is pure if it is not mixed with
pain or if it is not (with high probability) going to be immediately followed by
pain. By 'extension' is meant the number of people who are affected by the
experience. One can here note that 'extension' could be given another interpretation that is not noted by Bentham. An experience could be described as
extended when it affects a great deal of a person's mental life. This is easiest
to illustrate in the case of bodily sensation. A person who is injured all over the
body can be said to have a pain that has a greater extension than that of a
person who has an injury at only one place.

A significant omission in Bentham's list of properties is a dimension of
evaluation. Bentham does not indicate that certain kinds of pleasure are more
valuable than others. It seems as if Bentham believes that the 'value' of his
pleasures and pains can be objectively measured along the scales that his
dimensions suggest. If this is correct, Bentham clearly distinguishes himself
from his pupil Mill who explicitly talked about more valuable or less valuable
experiences. Mill's distinction between 'happiness' (the more valued experience) and 'content' (the less valued experience) is a crucial addition to the utilitarian philosophy.

According to Bentham the value of an experience of pleasure should be measured objectively, by assessing the experience according to the scales (referred to above). These could in principle, Bentham claims, be given cardinal values. Bentham then presents his famous 'felicific calculus'. Hereby he refers to the method by which one should be able to assess whether one piece of happiness is greater than another one. I quote extensively from Bentham (1982, p. 39):

> To take an exact account then of the general tendency of any act, by which the interests of a community are affected, proceed as follows. Begin with any one person of whose interests seem most immediately to be affected by it: and take an account,
> **1.** Of the value of each distinguishable pleasure which appears to be produced by it in the first instance.
> **2.** Of the value of each pain which appears to be produced be it in the first instance.
> **3.** Of the value of each pleasure which appears to be produced after the first. This constitutes the fecundity of the first pleasure and the impurity of the first pain.
> **4.** Of the value of each pain which appears to be produced by it after the first. This constitutes the fecundity of the first pain and the impurity of the first pleasure.
> **5.** Sum up all the values of all pleasures on the one side and those of all pains on the other. The balance, if it be on the side of the pleasure, will give the good tendency of the act upon the whole, with respect to the interests of that individual person; if on the side of the pain, the bad tendency of it upon the whole.
> **6.** Take an account of the number of persons whose interests appear to be concerned; and repeat the above process with respect to each. Sum up the numbers expressive of the degrees of the good tendency, which the act has, with respect to each individual, in regard to whom the tendency of it is good upon the whole: do this again with respect to each individual, in regard to whom the tendency of it is bad upon the whole. Take the balance; which, if on the side of pleasure, will give the general good tendency of the act, with respect to the total number or community of individuals concerned; if on the side of the pain, the general evil tendency, with respect to the same community.

Bentham then proceeds to make a detailed analysis and classification of the species of pleasure or happiness. This taxonomic enterprise gives a good picture of the broad area that Bentham has in mind. He first makes the distinction between simple and complex experiences. A single experience can be composed of experiences of different kinds. Here I list his main categories of pleasure.

1. Sense.
2. Wealth.
3. Skill.
4. Amity.
5. A good name.
6. Power.
7. Piety.

8. Benevolence.
9. Malevolence.
10. Memory.
11. Imagination.
12. Expectation.
13. The pleasures dependent on association.
14. Relief.

Some of the categories in this list require explication. By the pleasures of amity Bentham refers to such pleasures as are consequences of one's conviction that one has a good relation to one's friends. It is surprising, however, that love is not explicitly mentioned as a subcategory of this or any other of the categories of pleasure. The purely sexual pleasure is a subcategory of the pleasures of the senses. The pleasure of benevolence is the species that follows from a conviction that all is well with a person one likes. The pleasures of memory and imagination are some kind of second-order entities. Bentham reminds us how we can be happy not only about such phenomena as are directly presented to us but also about such things as we can only remember or just imagine.

The pains are given a similarly complicated taxonomy in Bentham's system. This on the whole corresponds to the taxonomy of happiness. As a real asymmetry in Bentham's system appears, however, one new category among the pains, i.e. the pains of privation. Bentham notes that we often feel pain because we lack a person or an object that we associate with pleasure. Knowledge about the existence of a pleasure that one lacks can give pain. Bentham distinguishes between three main categories of such pain: the case when the pleasure in question is strongly desired; the case when it is both strongly desired and expected and the subject therefore has become disappointed; and the case of sorrow or despair when the lack is associated with the knowledge that one will never more reach the desired person or object.

A Commentary on Bentham's System

A crucial distinction exists which Bentham does not make. This is a distinction that is a powerful tool for mapping the whole area of happiness and pleasure. I will return to it in a more systematic way when I introduce my own notion of happiness.

A first step in understanding this distinction is to think in terms of *direct* and *indirect* happiness or pleasure. What Bentham calls the pleasure of the senses is the paradigm of direct pleasure. By 'direct' I here mean that a person perceives an experience as pleasant without having to reflect upon the quality of the experience, upon its sources or upon its conceivable consequences. The pleasure of taste and the pleasure of sexual orgasm are direct in this sense. Most of the other species of happiness or pleasure mentioned by Bentham, and in general most of the other kinds that there are, are indirect in the sense that they are based on a *belief* or a *conviction* of their bearer. The pleasure of wealth, for instance, is dependent on the belief that one is well-off. The same is true of the pleasures of friendship, benevolence and malevolence. One must here be convinced that certain things are true in the world.

The pleasure (or pain) of imagination is an interesting special case. Here it is not a case of belief or conviction in the ordinary sense. Most often the imaginative person is conscious of the fact that he or she is imagining. This does not prevent there being a feeling of satisfaction at things that go well in the world of imagination. Imagination has a crucial feature in common with belief and conviction, however, viz. that it is cognitive.

Instead of using the terms 'direct' and 'indirect' pleasure we can now more adequately talk about sensual and cognitive pleasure. In my own proposal for a theory of welfare (Chapter 22) I will mark this distinction terminologically. I will call the former 'pleasure' and the latter 'happiness'.

5 A Background to the Analysis of Welfare or Quality of Life in the Human Sciences

Introduction

Having offered a brief historical introduction I now wish to turn to the modern discussion of welfare concepts with regard to humans. A first observation is that the term 'welfare' is hardly ever used today in the human health-care sector. 'Welfare' has a central place in sociological theory, where it is almost universally used as a term referring to social resources: institutions, insurance systems, provision of housing, vital food resources, etc. The terms used to characterize a state of an individual human being are normally 'well-being' or, more frequently, 'quality of life'. I will in the following focus on the latter expression and its possible content.

An Analysis of the Concept of Quality of Life in the Human Context

The expression 'quality of life' is composed of two terms: 'quality' and 'life'. Both are in need of analysis. Notwithstanding the increasingly rich literature on quality of life, the basic notion of life has been almost completely neglected in this context. I will here make some observations.

An important distinction is that between a complete life and a partial life. Moreover there are at least two dimensions along which the degree of completeness can be measured. One dimension has to do with time; another has to do with aspects of life. A complete life in the former sense is composed of the continuous series of life events that a particular person goes through during his or her existence from birth to death. A complete life in the latter sense is the sum total of all aspects of his or her existence at a certain moment or during a certain period. A maximally complete life is then the sum total of all the aspects of a person's existence during his or her entire lifetime.

At least the following main aspects of life could be considered:

1. The experiential aspect of life; the sum total of a person's sensations, perceptions, cognitions, emotions and moods.
2. The activities in life; the sum total of a person's actions.
3. The achievements in life; the sum total of the results of a person's actions.
4. The events in a person's life, those that the subject is aware of or which are ascribed to him or her, or both.
5. The circumstances surrounding a person, either such as the subject is aware of or such as are ascribed to him or her, or both.

Different theories of quality of life have focused on different aspects of life. An important further distinction is between, on the one hand, such aspects of a person's life as could be objectively ascribed to him or her, on the other hand such aspects as the subject perceives or otherwise experiences. This is one of several senses in which we can talk of objective quality of life and subjective quality of life.

In the discussion with regard to human well-being and quality of life it is often assumed that what constitutes well-being or quality of life is determined in an evaluative decision. We *choose* what constitutes the good life. (The term 'goodness' clearly indicates the evaluative nature.) The choice is not necessarily conscious. A person, when explicitly asked, may make a mistake about his or her 'deep-down' evaluation. Gaining access to this deep-down evaluation is difficult. It may emerge in the form of happiness, but then we must consider a person's well-being and quality of life in the long run. (See further discussion below in this section.)

However, we can choose to evaluate a human being's quality of life along more than one evaluative dimension. There are for instance moral values, prudential values, aesthetic values, intellectual values, values of decency and values of welfare.

A crucial question in quality-of-life research is: who should be the evaluator of a particular person's life? According to a Platonic–Aristotelian conception the answer would run as follows. There are objective values. The evaluations are given once and for all. There is no room for personal choice among values. As a result, it does not matter – apart from the possibility of making mistakes – who exercises an evaluation in a particular case. As long as the person making the evaluation has the proper insight into the realm of absolute values, then this person will also come up with the correct evaluation of the quality of life of the person in question. In particular, Aristotle gave us a substantial treatment of the perfect life and he suggested that the best life, all things considered, is the life in accordance with the highest of virtues. Such a life is what he called the *eudaimonian* life (see Chapter 4).

There is no single contemporary view of these matters. There are contemporary ideas that come close to the Aristotelian one (Nussbaum and Sen, 1993). But there are contemporary views that are its direct opposite

(Nordenfelt, 1987/1995; Sandøe and Kappell, 1994; Veenhoven, 2000). Moreover, there are some in between (Naess, 1987; Kajandi, 1994; Brülde, 1998). The anti-Aristotelian view entails that there is no hierarchy of values given once and for all. All evaluations are made by individuals and therefore every assessment is dependent on the person who makes the evaluation.

This is the problematic starting-point for the empirical science of quality-of-life research and, I would argue, also for animal welfare research. What dimensions of evaluations and what criteria should we follow? In the human case there seem to be two kinds of plausible strategies available; one more collective and paternalistic, the other more liberal and individualistic. The former strategy, which could be realized in a variety of ways, would entail the following. A number of 'experts' or politicians come together to decide what is the essence of quality of life (sometimes shortened QoL). They decide first what aspects of life are the most important ones for the particular purpose chosen. Second, through consensus reasoning or through a simple majority vote they decide on the scale along which individual lives should be measured.

This could be done in a more or less *a priori* fashion. The people involved could have an Aristotelian view and attempt to work out the details of their *eudaimonia* concept and then try to establish this as the basis for assessment. (An example of this is perhaps the need approach for the evaluation of quality of life. For an analysis see Chapter 15.) A more *a posteriori* (or democratic) procedure would involve making an empirical investigation and obtaining an idea about how people in general evaluate their lives. This is the procedure chosen for the establishment of some measurement scales in health care.

The individualist strategy entails that there is no general instrument for the assessment of quality of life. The person who has the task of making the assessment, whether this is a physician or a social worker, permits the subjects themselves to make the evaluation. The subjects can then make their life evaluation according to their own preferences. X's quality of life becomes identical with X's own evaluation of his or her quality of life.

But how shall we understand the locution: X's evaluation of his or her quality of life? And how can we come to understand this evaluation? There are at least two crucial understandings of the locution, one more superficial than the other, and they have methodological implications. The most straightforward and superficial interpretation is: X's evaluation of X's quality of life is identical with X's explicit answer to a question concerning X's quality of life, presupposing that X has correctly understood the question and attempts to answer honestly.

The other interpretation is the following: X's evaluation of his or her quality of life is the evaluation that X would make given that he or she could organize his or her value system coherently and could give a proper assessment of his or her life according to this system.

The latter is, then, the ideal notion of subjective evaluation of quality of life. No subject can fully accomplish this, however. Most of us lack coherent value systems and we are unable to remember and truly grasp all the relevant facts about our lives. Some individuals are particularly poor at this task. They may not have intellectual qualifications. Some human beings, not only infants and mentally retarded people, do not understand the notion of a value.

Thus, to achieve a deep subjective evaluation of a life seems to be an extremely difficult task. On the other hand, this interpretation seems to be the more reasonable. A person's evaluation of his or her quality of life need not be identical with what he or she 'honestly' says in reply to a question. If it were, then there would be no room for a subject to make a mistake in assessing his or her quality of life.

But if hardly anybody can achieve full insight into his or her quality of life even on this individualistic interpretation, what are the methodological implications? Here, some theorists seem to propose a compromise.

1. There is good reason to believe that a normal adult is capable of providing a rough estimate of his or her quality of life. Thus an ordinary and well-designed questionnaire might be sufficient.

2. Some people – infants, the mentally retarded, some psychotics and the demented – have no or little capacity to express their view about quality of life. In such cases other people must assist and do the evaluation for them. Still, this can be done from the individual's point of view. It remains the subject's quality of life that should be assessed and no one else's.

6

Some Contemporary Theories of Human Welfare or Quality of Life

A General Theory of the Good Life: Bengt Brülde

The Swedish philosopher Bengt Brülde (1998) has proposed a sophisticated theory of welfare or the good life that has borrowed elements from both an Aristotelian view and a utilitarian/hedonistic one. It also contains an essential element dealing with preference satisfaction. According to Brülde there are first two kinds of situations that are good for a person: (i) to have certain kinds of pleasant experiences (cf. Bentham); and (ii) to have his or her relevant desires fulfilled. In addition there are some objective requirements: it is not good for a person to take pleasure in something 'negative', and it is not good for a person to have a desire fulfilled which is on the 'negative' list (cf. Aristotle).

Brülde does not offer a complete account of what should count as the 'objectively' good values. (He rejects, however, a number of traditional such accounts.) An example of what he means is the following. It is better to take pleasure in something that is true or authentic than in something that is false or inauthentic. Similarly it is better to have a desire satisfied when this desire is worth desiring. A clear example would be: it is better to have one's desire for rest fulfilled than to have one's desire for alcohol intoxication fulfilled.

But how do we determine, according to Brülde, how good a certain situation is for a certain person? In the case of valuable pleasures, the value that it has for a person to have such an experience is normally a function of how pleasant the experience is, e.g. it is better to have a more pleasant experience than to have a less pleasant one. But sometimes the value is also, as mentioned above, dependent on whether the pleasure is based on true or false beliefs. In the case of desire-fulfilments, the value it has for a person to have a relevant desire fulfilled is normally a function of how strong the desire is, i.e. it is better to have a stronger desire fulfilled than a weaker desire fulfilled. But sometimes other considerations can come in. It is better, *ceteris paribus*, for a person to have a desire fulfilled if its object is among the 'objectively' good values than if it is not.

A person's well-being is thus according to Brülde roughly a function of how much valuable pleasure and how much valuable desire fulfilment there is in the person's life. This idea can be simply formulated in terms of 'happiness', if we make the following assumption. 'Happiness' often refers to pleasant feelings but it can also mean that the subject endorses a particular fact, i.e. there is both an affective and an attitudinal component in the concept of happiness. Then we can say: a person's level of well-being is (roughly) a function of how happy he or she is with life, but only on the assumption that the affective component is based on true beliefs regarding what life is like.

To this must be added a list of objective factors. Brülde summarizes this requirement in the following way:

> A person's level of well-being is (roughly) a function of how satisfied he is with his existence, but only as long as he takes all the relevant dimensions into account. Or more specifically, the satisfaction which determines how well off a person is (on the whole) must (so to speak) 'include' how satisfied he is in a number of 'objectively predetermined' areas, and a person's level of satisfaction in the relevant areas must (roughly speaking) be 'in line with' the different objective values.
>
> (Brülde, 1998, p. 377)

One can conclude that if this theory is to be of pragmatic use, then a thorough analysis of what should count as relevant areas and objective values is urgently needed.

Siri Naess and Quality of Life: a Psychologist's Notion

Let me now turn to some more pragmatic theories of quality of life. These theories are often designed for a specific purpose, for instance the assessment of people's well-being in a social context or in the context of health care. To my knowledge there is in these discussions no purely Aristotelian notion of quality of life expressed. Welfare or quality of life, according to most contemporary social and medical theories, includes a psychological component of subjective well-being, although it frequently also contains other elements such as function and activity.

A good representative of a contemporary psychological theory of quality of life is the Norwegian scholar Siri Naess (1987). I mention her since she has been quite influential in Scandinavian attempts to construct measures of quality of life. According to Naess's theory quality of life has four essential components: (i) activity; (ii) good interpersonal relations; (iii) self-esteem; and (iv) a basic mood of happiness. These central components can in their turn be divided into a number of components:

Activity	⎧ Enmeshment ⎨ Energy ⎩ Self-realization Freedom
Interpersonal relations	⎧ Intimate relation ⎨ Friendship
Self-esteem	⎧ Self-confidence ⎨ Self-acceptance
Basic mood of happiness	⎧ Emotional experience ⎨ Security ⎩ Happiness

Naess claims that the main components, as well as the sub-items, are equally important parts of one's quality of life. No category can be reduced to another. This is not to deny that there may be empirical relations between items in the different categories. It is quite likely, for instance, that items in the first three categories influence items in the fourth. One may therefore be tempted to say that the real well-being lies in category 4, and that the others are presuppositions of 4 (i.e. playing the role of sources or conditions of happiness).

This is not, however, Naess's contention. Category 4 is not primarily a category for such happiness as has items in categories 1, 2 and 3 as its object. Instead, category 4 comprises a mixed bag of character traits, cognitions, moods and general happiness with life. Of importance is that most of Naess's components refer to intra-personal categories. Quality of life with Naess is basically a psychological concept. What is to be assessed and measured is a person's subjective state and not his or her objective surroundings.

Although this is the main intention there are such items as presuppose features in the external world; in particular some of the items refer to relations between the subject and (normally) other people. Under the heading of activity one finds an item saying that the subject has opportunities to fulfil his or her plans and can take part in decisions in a social context. Category 2, Interpersonal relations, is basically a category of relations, although it is sometimes inserted that the subject evaluates these relations. Relations, such as having the opportunity to realize plans or being a member of a professional group, are compatible with the fact that the opportunity is not used or that the membership does not lead to a feeling of contact.

This indicates that Naess's concept of quality of life is not an exclusively hedonistic concept. The good life for her requires that certain objective facts hold. The lonely person, the person who does not belong to any groups or has few opportunities in life, is, per definition for Naess, a person with a reduced quality of life. She says: 'It is better to be happy together with others than to be happy alone. If person A and person B are equally active, happy and have an equally high self-esteem, but A has better social relations, we would consider A's quality of life as being higher per definition' (Naess, 1987, p. 18). The main reason why Naess incorporates such *eudaimonian* elements in her concept of quality of life is her reluctance to equate quality of life with pure satisfaction of wants. (In Nordenfelt, 1993, I have defended a want-satisfaction theory and responded to some of Naess's arguments against it.)

A Swedish psychologist, Madis Kajandi (1994), has constructed an instrument for the measurement of quality of life based on Naess's conception. He stresses even more the idea that the concept of quality of life contains objective elements. He divides the concept into: (i) external conditions of life; (ii) interpersonal relations; and (iii) inner mental states. As paradigm cases of external conditions Kajandi mentions working situation, economic situation and living situation. Interpersonal relations are divided into pair-relations, relations of friendship and family relations. Inner mental states divide into enmeshment, energy, self-realization, freedom, self-confidence, self-acceptance and emotional experiences.

Veenhoven and the Four Qualities of Life

The Dutch psychologist Ruut Veenhoven (2000) has made some valuable distinctions within the area of quality of life, which he has summarized in the following matrix:

	Outer qualities	Inner qualities
Life chances	Livability of environment	Life-ability of the person
Life results	Utility of life	Appreciation of life

In the upper half of the scheme are presented two variants of potential quality of life, says Veenhoven. The top left quadrant concerns good living conditions. This is the area sometimes called welfare by economists. (See also my own proposal in Chapter 22.) The top right quadrant concerns inner life chances. By this Veenhoven means how well we are equipped to cope with the problems of life. This is also roughly what Amartya Sen (1992) calls capabilities (see Appendix in this book). The lower half of the scheme is about quality of life with respect to its outcomes. These outcomes can be judged by their value for one's environment and the value for oneself. The external worth is called 'utility of life' and the internal worth is called 'appreciation of life'. The latter is also commonly referred to by terms such as 'life-satisfaction' and 'happiness'.

Veenhoven considers in detail what possibilities we have to assess and measure the four qualities of life. His general conclusion is that measures of livability, life-ability and utility of life are either non-existent or have serious limitations. However, Veenhoven contends, it is much easier to measure the subjective enjoyment of life. Since this is something people are directly aware of we can simply ask them, and this is now what is often done in questionnaires and interviews in social science and medicine (see further below).

An advantage with measuring subjective well-being or life-satisfaction is also that it captures well at least three of the four qualities of life. Subjective well-being implies two things: first that the minimal conditions for the person's thriving are met; and secondly that the fit between opportunities and capacities must be sufficient. Hence subjective well-being, while a kind of quality of life in itself, can function as a summary of two of the others as well. However, Veenhoven contends, a life can be happy but not useful, and it can be useful but not happy.

Veenhoven's main conclusion is that the most comprehensive measure for quality of life is how long and happily a person lives. At the same time he acknowledges the existence of the other qualities. His own empirical research, however, is completely geared to the assessment of subjective well-being.

Health-related Quality of Life: Medical Instruments

A problem for the analysis of notions such as health, welfare and quality of life is that the concepts in many systems tend to overlap. This is most easily seen in the medical context. Here instruments are constructed for the measurement of quality of life, but sometimes the same instruments are referred to as health measurement instruments. A further alternative is to categorize these instruments as measuring health-related quality of life. I will argue that the last expression is often the most adequate one.

The typical assessment of quality of life in the medical context serves practical purposes. One such purpose is the testing of a new medical methodology, for instance a drug that affects quite a limited bodily function, or that has only a pain-relieving effect. Another concrete use is when the purpose is directly to improve a person's quality of life. Here, the task is oriented to the person's present situation in quite a narrow sense. Moreover, the medical assessors of quality of life tend to select very carefully among the properties of the person to be assessed. They concentrate on certain bodily and mental, mainly experiential, aspects. Normally, however, a set of circumstantial facts is also included.

To get an overview of quality-of-life instruments in medicine one needs to distinguish between psychiatry and other medical disciplines. The psychiatric instruments tend to cover more aspects of a human life. In fact, the instrument designed by Kajandi above is made for psychiatric purposes. The somatic ones, on the other hand, can be very limited in content and may only mirror such aspects as are closely related to the subject's somatic health.

A Classical Measure of Health-related Quality of Life: the Nottingham Health Profile

One of the oldest and, at least in Europe, most frequently used instruments for the measurement of health-related quality of life is the Nottingham Health Profile (NHP) (Hunt, 1988). The NHP is basically a questionnaire divided into two main sections. The first section contains 38 statements related to a person's experience of health and illness. The subject is supposed to answer yes or no to each of the statements. The second section asks in general whether the respondent's state of health is causing him or her problems with various parts of his or her life, such as work, social life, home life and sex life.

The first, more detailed, section can in its turn, from a semantic point of view, be divided into six categories. These categories are given the following labels: physical mobility, pain, energy, sleep, emotional reaction and social isolation. Some statements in the questionnaire refer to quite serious states such as: I feel that life is not worth living; others are quite moderate: I find it hard to reach for things.

The severity of a state of affairs plays a role in the final evaluation of the person's health-related quality of life. The statements are ranked as to severity within each category. The ranking was not derived from a logical analysis of the statements. Nor does the individual whose quality of life is assessed perform his or her own ranking. Instead a population of several hundred patients and non-patients were asked to make their judgements of the severity of the states referred to.

A major problem with the NHP, a problem that it shares with several instruments for measuring health-related quality of life, is that some of the variables are not necessarily tied to the subject's health status. Consider, in particular, the categories of sleep, social isolation and emotional reaction. Almost all the items in these categories are compatible with an experience of good health, as intuitively understood. We can imagine a healthy person placed in a difficult and tragic situation that has nothing to do with his or her body or mind. The person may have lost a close relative, may have lost his or her job, or may be faced with extreme financial difficulties. Confront then such a person with the following statements:

I lie awake for most of the night.
Worry is keeping me awake at night.
Things are getting me down.
I lose my temper easily these days.
The days seem to drag.
I am finding it hard to get on with people.

It requires little imagination to see that these mental states are adequate reactions to a great variety of unhappy circumstances of which illness is only one. The items are open in the sense that there is no reference to the objects or causes of the mental states. The situation would have been quite different with the following phrasing: I sleep badly *because of* my illness, or I lose my temper easily *because of* my illness.

A European Project for Measuring Health-related Quality of Life (EuroQol)

An attempt to measure health-related quality of life, similar to the NHP project, is called EuroQol (Brooks, 1996). The concept of EuroQol and its instrument have attracted a lot of interest among persons involved in the evaluation of medical technologies. The EuroQol questionnaire has been translated into several languages and is being used all over Europe. Like NHP its constructors intend it to serve a general purpose. EuroQol is a questionnaire not attached to any specific group of diseases. It should be able to measure a person's general health or health-related quality of life. (The EuroQol constructors normally use the term 'health' although the items in the instrument need not be health-related.)

The dimensions covered by EuroQol are the following: mobility, hygiene, main activities (such as work, studies or household activities), pain/discomfort, and anxiety/depression. With regard to each dimension the respondent is supposed to state whether he or she has great problems, moderate problems or no problems at all. The five dimensions and the three levels allow 243 combinations, i.e. 243 possible health states. These states (or in practice a subset of 42 states) were assessed by a selection of the general British public in a study carried out in 1993 (described in detail in Dolan *et al.*, 1995 and Williams, 1995). The method used for assessment is called Time Trade-Off. This method entails an investigation where life-quality is assessed in relation to duration of life. The subject is asked how many years of living in health state X is equivalent to one year in a completely healthy state. Assume that a person has a choice between, on the one hand, living one year as completely healthy and, on the other hand, living 9 years as having moderate problems along all the five dimensions and finds that these two choices are equivalent. Then this relation (1 to 9 years) can be used for the calculation of the utility value for this specific health state.

The EuroQol concept faces the same methodological problems as the NHP (although the methods of EuroQol are in many ways more sophisticated), viz. that the final evaluation of a health state is based on the judgements of a particular selection of persons in a specific culture such as the British culture. A comparative study has been made where the same assessments were made by a group of Swedes. The result was that with regard to certain health states (in particular in the middle range of the scale) there were great differences between the judgements of the citizens of the two countries (Henriksson and Carlsson, 2002).

EuroQol also shares with the NHP the crucial property that some of its items, in particular anxiety and depression, can be the result of other phenomena than disease or injury. A person in a catastrophic situation, for instance in a country at war, can be in a state of great anxiety and a person who has lost his or her nearest and dearest can be in a state of greatest depression. Anxiety and depression are indeed relevant for the measurement of general quality of life but not necessarily for the measurement of health-related quality of life.

II Theories of Animal Health and Welfare

7 Ideas on Animal Health

An Analysis of Textbooks in Veterinary Medicine

If the concept of human health is central in many scientific debates this is not so with animal health. Although animal health is central to veterinary medicine there are hardly any fora where this concept has been subjected to scrutiny. The reasonable diagnosis seems to be that the concept is not viewed by the professionals as problematic.

This conclusion is also drawn by the Swedish veterinarian Stefan Gunnarsson (2005), who has made a survey of veterinary textbooks with regard to their treatment of the concepts of health and disease. Gunnarsson consulted 500 modern textbooks in animal pathology, epidemiology, internal medicine and some other areas. Only 39 of these books (8%) contained any explicit definition of health and/or disease. Twenty-two of the latter were written for veterinarians or veterinary students. Five were dictionaries, two were handbooks for veterinary nurses and ten were written for farmers or animal owners. Most (25) of the books (in the 8% category) were in English, one was in French and the others were in one of the Scandinavian languages.

It was striking that fewer of the textbooks directed to the veterinarians or veterinary students contained definitions of health or disease than did the others. There were, for instance, three books on alternative veterinary medicine written for lay people where there were explicit definitions of health.

Gunnarsson finds quite a variety of health definitions, comparable in fact to existing definitions in the human field. He makes the following tentative subdivision:

1. Health as normality.
2. Health as biological function.
3. Health as homeostasis.

4. Health as physical and psychological well-being.
5. Health as productivity, including reproduction.

The first two categories seem to overlap, if one scrutinizes the examples Gunnarsson provides. This is natural since a standard theory of health is in terms of *normal biological function* (see Boorse, 1997, above). In some cases, however, only the normality is stressed, for instance 'normality in posture, movement, alertness and appetite' (Webster, 1987). Another author, Cheville (1988), underlines that 'pathology in the broadest sense is abnormal biology'.

The old idea of health as balance or homeostasis recurs in some of the textbooks scrutinized. It is particularly salient in the paper by Day (1995), who writes about homeopathic veterinary medicine: 'Like all systems in equilibrium, the body – a very sensitive and active equilibrium system – reacts to disturbing forces in an attempt to retain or regain balance'.

With the heading 'physical and psychological well-being' Gunnarsson refers to such theories as have an inclusive health concept, embracing the psychological and even the social sphere. Here he mentions such definitions as align themselves with the famous WHO characterization: 'health is complete physical, mental and social well-being'. Blood and Studdert (1999) combine this idea with the health concept that is the final one in Gunnarsson's list by saying: 'health is a state of physical and psychological well-being and of productivity including reproduction'. Grunsell (1995) in *Black's Veterinary Dictionary* states, however, more bluntly: 'health is now more accurately regarded as a state of maximum economic production'.

Thus the general picture is that only a few veterinary textbooks include any explicit definitions of health and disease. This is, indeed, not different from the case in medical textbooks. A major difference, however, is that in some sub-disciplines within human medicine and human health sciences there is some systematic discussion about concepts of health, illness and disease. Moreover, other disciplines such as anthropology, sociology and philosophy have entered the arena and put the notions of human health, disease and illness under scrutiny. Such a development does not yet exist in the case of veterinary medicine.

Donald Broom's Theory of Health

The British animal scientist Donald Broom is, however, an exception to the claim that there are virtually no explicit characterizations of health in the animal science literature. I quote: 'The word "health" like "welfare" can be qualified by "good" or "poor" and varies over a range. However, health refers to the state of body systems, including those in the brain, which combat pathogens, tissue damage, or physiological disorder'. He adds, in order to distinguish health from welfare: 'Welfare is a broader term covering all aspects of coping with the environment and taking account of a wider range of feelings and other coping mechanisms than those that affect health, especially at the positive end of the scale' (Broom, 1998, pp. 395–396). Although this description tells us something it is not so informative about the nature of health.

More informative, and running in the same direction, is the discussion in Broom and Kirkden (2004). Broom there introduces an explicit definition of health: 'Health may be defined as "an animal's state as regards its attempts to cope with pathology"' (p. 341). Broom (2001, p. 4) again adds the crucial specification that health is a part of welfare. Thus it is not correct to say 'consider the health and welfare'. One should rather say: 'consider the welfare including the health'. Pathology in its turn is defined as a detrimental derangement of molecules, cells, tissues and functions that occurs in living organisms in response to injurious agents or deprivations. Pathology, according to Broom and Kirkden, can be classified into: genetic abnormalities; physical, thermal and chemical injury; infections and infestations; metabolic abnormalities; and nutritional disorders (p. 342). It is striking, though, that there is no talk of mental pathology here.

It is clear first that the scope of health for Broom is smaller than the scope of welfare. When we say that a person is healthy we are, according to him, only referring to the state of a part of his or her defence system. More precisely, we are referring to the parts of the body (and mind?) that combat 'pathogens, tissue damage, or physiological disorder'. This, then, is quite a narrow interpretation of health.

Consider the following possible interpretation of Broom's characterization:

1. An animal is healthy if, and only if, its immune and, in general, reparative systems are in order.

Given such an interpretation an animal can be healthy during an infection. As long as the system for combating infection is working as it is designed to do, the animal is healthy in spite of the presence of a disease.

However, a different and more plausible interpretation is the following somewhat wider one:

2. An animal is healthy if, and only if, the immune and reparative systems actually succeed in eliminating pathology, i.e. if the animal as a result is without disease (induced by pathogens), tissue damage or physiological disorder.

This is a common idea also in the literature with regard to human health. Health is the absence of disease, injury and defect. Boorse (1997), as we have seen, proposes such a definition and it seems reasonable to interpret Broom as an adherent of this view. The task we encounter then is to define the negative concepts: disease, injury, etc. A fundamental part of Boorse's project is to propose such definitions. His suggestion is that a disease is a state of the body causing or entailing a reduced function in relation to the biological goals of survival and reproduction. Completely parallel characterizations can be given for injuries and defects. The healthy human or animal then becomes the one where all parts function properly in relation to survival and reproduction.

This general idea can be extended to cover also mental health. The mental faculties express themselves in terms of behaviour. The mentally ill person often behaves in ways that are not fruitful for his or her survival or reproduction and these behaviours are often included in the very definitions of the illness. Thus, several mental illness concepts entail that there is a reduction in the individual's behaviour in relation to the biological goals of survival and reproduction.

But does this extended biostatistical theory of health then entail that a person is in good health whenever his or her body and mind are *coping* well? We shall see that the idea of coping is central among animal scientists, but then primarily in relation to the general notion of *welfare*. This holds in particular for Broom. A crucial task for us will therefore be to consider the similarities and differences between health and welfare according to the theory of Broom and other animal welfare scientists. I will return to this issue in Chapter 9.

Broom and Kirkden (2004, p. 344) note that some veterinarians would define animal health more broadly than what is suggested above. They mention Blood and Studdert's (1999) definition which says that health is a 'state of physical and psychological well-being and of productivity including reproduction'. Broom and Kirkden claim that this definition is inadequate, 'partly because well-being is not defined and partly it is far too inclusive. It does not reflect the practice of veterinary medicine, which is primarily concerned with physical abnormalities'. We can observe that Broom and Kirkden (like Boorse on the human side) are not prepared to accept an idea of positive health released from the notions of disease and injury.

A Holistic Notion of Health in Veterinary Science

In contrast, a recent article (Hovi *et al.*, 2004) proposes a completely different approach with regard to animal health. The authors take as their starting-point the WHO definition of health from 1948: 'Health is a state of complete physical, mental and social well-being and not merely the absence of disease'. Hovi *et al.* sympathize with the idea of a holistic approach to the characterization of health. They think that modern veterinary medicine must embrace a broad concept covering physical, emotional and mental levels in general. According to their idea health is 'a characteristic of a living individual and can be understood as an expression of harmony or balance in the individual at all levels' (Hovi *et al.*, 2004, p. 255).

Hovi *et al.* emphasize that if one concentrates on health rather than disease the focus is moved from the diagnosis of disease and identification of risk factors towards health promotion. One could then, they say, primarily deal with positive health planning and downplay conventional disease prevention involving merely the identification and removal of risk factors. In the latter procedures the disease is in focus, whereas according to a positive (in the article also called ecological) concept of health the entire animal and its surroundings are in focus.

The article by Hovi *et al.* (2004) explores further the notion of health planning. The analysis is partly based on the UK national organic standards from 2001, where an animal health plan is explicitly required:

> The plan must ensure the development of a pattern of health building and disease control measures appropriate to the particular circumstances of the individual farm and allow for the evolution of a farming system progressively less dependent on allopathic veterinary medicinal products.
>
> (Hovi *et al.*, 2004, p. 258)

8 Some Examples of Ideas on Animal Welfare

One can find a great variety of ideas on animal welfare in the contemporary literature. There is first a set of theories that are *biologically oriented* and more or less inspired by evolutionary theory. This orientation is most clearly seen in C.J. Barnard and J.L. Hurst (1996): 'Welfare can be interpreted only in terms of what natural selection has designed an organism to do'. A similar idea is found in John McGlone (1993): 'I suggest that an animal is in a state of poor welfare only when physiological systems are disturbed to the point that survival or reproduction are impaired'.

The biological approach is fundamental also in Donald Broom's theory (1986 and several other works): 'Welfare of an individual is its state as regards its attempts to cope with its environment'. Broom has, however, modified his position into a more complex theory involving feelings in his later writings.

A strong group of theories argue for a concept of welfare in terms of *feelings of well-being*. Ian Duncan (1993) is an exponent of this view: 'Neither health nor lack of stress nor fitness is necessary and/or sufficient to conclude that an animal has good welfare. Welfare is dependent on what animals feel'. His British colleague Marian Stamp Dawkins (1990) expresses a similar view. The idea that all higher animals are sentient beings, and thus can suffer, lies at the bottom of the ethical arguments for animal welfare and thereby at the bottom of the political animal welfare movement.

A third group of theories centre around the ideas of animal welfare as *fulfilment of needs, preferences or wants*. Here S.E. Curtis (1987) can serve as an example: 'Welfare of an individual is its state where it can fulfil its needs/wants'.

A fourth group of theories focus on the idea of *natural behaviour*. An animal has welfare, according to this idea, when the animal can behave in its natural way, unimpeded and not forced to behave in a certain direction. B.E.

Rollin (1992) is a forceful exponent of this reasoning: 'Welfare of an animal is its freedom to perform most types of natural behaviour'.

A fifth standpoint is the one represented by David Fraser, where he proposes that welfare should be analysed as a *complex concept* including the elements of natural behaviour, affective experiences and normal biological function. Fraser *et al.* (1997) note the following three fundamental ethical concerns in the animal welfare literature. 'First are natural living concerns which emphasize the naturalness of the circumstances in which animals are kept and the ability of an animal to live according to its "nature"' (p. 190). A second type of concern emphasizes the affective experiences of animals. The good life of the animal is thought to depend on freedom from suffering in the sense of pain, fear, hunger and other negative states of feeling. 'Thirdly, functioning-based concerns, held especially by many farmers, veterinarians and others with practical responsibility for animal care, accord special importance to health and the "normal" or "satisfactory" functioning of the animal's biological system' (p. 191).

Finally, one can find theorists who underline the *closeness between the notions of animal health and animal welfare*. Hughes and Curtis (1997, p. 109) express themselves in the following way: 'The relationship between health and welfare is self-evident. Some people regard them as almost synonymous'. Even more common is to say that welfare, if not the same as health, at least includes health.

9   Biological Theories of Animal Welfare

An Evolutionary Theory of Welfare

Among the biologically oriented theories of welfare a purely evolutionary perspective is taken by C.J. Barnard and J.L. Hurst (1996). Their theory of welfare completely excludes notions of subjective well-being and suffering and it even denies the scientific viability of such notions. They contend that welfare can be interpreted only in terms of what natural selection has designed an organism to do and how circumstances impinge on its functional design. In order to assess a state of welfare we should try to understand the organism's naturally selected performance criteria.

The authors are particularly sceptical as to analyses of welfare in terms of feeling states. They doubt that we shall ever have direct and unequivocal means of appreciating the subjective experiences of other organisms, even our conspecifics. 'Inferences about negative subjective states in others that we might wish to compare with suffering can therefore be made only on some "benefit of the doubt" basis' (Barnard and Hurst, 1996, p. 408). (For completely opposite views, see Chapter 12.)

Barnard and Hurst are also sceptical, however, with regard to the idea of equating welfare with coping à la Donald Broom (see the next section). They talk about the fallacy of individual preservationism. It is plausible to believe, they say, that there is a degree of self-preservation in the strategies of all organisms. However, there may be other strategies that may come into conflict with self-preservation and which are necessary for the individual's reproductive success. Physical impairment or even death may be adaptive trade-offs favoured by selection:

> From an evolutionary viewpoint ... individuals are expendable commodities in the pursuit of reproductive success ... On this view, measurable costs such as

reduced growth, increased fluctuating asymmetry, immuno-depression, pathology and injury may reflect adaptive trade-offs that the organism is designed to accept. While the organism may experience negative subjective states associated with these costs (e.g. fatigue, pain, hunger, nausea, fear), they are part of the mechanism of naturally selected regulatory processes that optimize its activities during its life-time rather than a reflection of circumstances in which it is not designed to be.

(Barnard and Hurst, 1996, pp. 410–411)

The organism's priority, according to Barnard and Hurst, is, then, to maximize reproductive success through efficient self-expenditure; the priority is not necessarily the preservation of the organism itself. Barnard and Hurst thus identify the *welfare* of animal A with the maximization of A's reproductive contribution. In many instances the behaviour and development leading to such success are equivalent with self-preservation and growth. In these cases welfare à la Barnard and Hurst might coincide with ideas of welfare taken from the human discourse. However, since natural selection may have designed an organism for suffering and self-expenditure in certain contexts, suffering and self-expenditure may in these contexts constitute a high level of welfare. Thus it is not at all self-evident that the (evolutionary) welfare of animals should consist in self-preservation or positive feelings in those animals where such can occur. In the extreme case 'the best thing' for A to do in order to maximize its reproductive contribution is to die. Its dying process could thus, paradoxically, constitute its welfare.

In his excellent critique of current interpretations of evolutionary theory, G.C. Williams (1966) makes a similar observation with regard to the relation between evolutionary fitness (in the Barnard and Hurst sense) and traditional understanding of individual health and well-being. Williams concludes (p. 26) that reproduction always requires some sacrifice of resources. An individual's well-being and health can clearly be jeopardized in its struggle for maximum reproduction.

It is interesting to see that this use of evolutionary theory for the construction of an idea of individual welfare (or health) is different from the one discussed in Chapter 3. For Wakefield the role of evolutionary theory is mainly to help us identify the natural functions of individuals. When the organs of an individual fulfil the natural functions, then the individual is healthy according to Wakefield. Barnard and Hurst propose something more complex. The individual has welfare, according to them, when it is maximally efficient in producing offspring, i.e. when it is fit in a strict evolutionary sense. On the other hand, they contend that welfare can only be interpreted in terms of what natural selection has designed an organism to do. These two interpretations can go together in many instances. But they need not. Great efficiency in producing offspring is clearly compatible with the fact that certain organs or certain behaviours of the individual are 'unnatural' or that they do not fulfil the 'natural' functions believed to be selected through evolution.

Thus Barnard and Hurst's position is not entirely clear. On the one hand, when they emphasize evolutionary fitness, they seem to be far away from the mainstream of contemporary concerns for animal welfare. How could this standpoint inform us with regard to the issues facing welfare theorists today? Would there be any place for scientific communication about the welfare of our pet dogs, our riding horses and the sheep and cows that are being transported on the roads of Europe? An extremely negative conclusion might be that there is no animal welfare over and above efficient reproduction. Other concerns, like the ones expressed and presented in official documents or in contemporary ethical discourse, would then have no scientific basis. On the other hand, when Barnard and Hurst talk about functional design and the impact of circumstances on such design, they approach the view of Wakefield. Here, welfare could in principle be ascertained without the consideration of individual reproductive success.

In their summarizing paragraphs Barnard and Hurst emphasize that they do not deny the existence of subjective feelings. Their main point is that we do not have any scientific means of assessing subjective feeling states. We can at most measure feelings in a proxy way via the identification of functional design. But one can wonder why Barnard and Hurst, given their basic philosophy of welfare, make concessions to feelings at all. If feelings were eventually found to play a minor role or even a negative role with regard to an organism's reproductive fitness, then they ought not have a place in the theory of welfare, according to Barnard and Hurst.

A Theory of Welfare in Animal Science in Terms of Coping (the Theory of Donald Broom)

In a number of papers from the 1980s Donald Broom has presented an influential biologically orientated theory (or, rather, different versions of a theory) of welfare with regard to animals. Broom's analysis almost exclusively concerns animals and concepts of animal science but he has suggested that his theory is applicable all over the animal world including humans. His claim, therefore, is similar to the one made by Christopher Boorse (see above, Chapter 2). Observe, however, that Broom talks about his theory of *welfare* as universal, whereas Boorse's theory is explicitly about *health*.

It is striking that Broom's conception of welfare for animals cuts through the corresponding conceptions of human health and quality of life. Much of what is said by Broom and others about welfare would instead be ascribed to health in the human case. Coping and fitness are central notions in Broom's conception of welfare. (The centrality of fitness is even more obvious in the case of John McGlone: 'I suggest that an animal is in a state of poor welfare only when physiological systems are disturbed to the point that survival or reproduction are impaired', 1993, p. 28.)

These notions are, as we have seen, central in several analyses of human health. It is therefore not only, as one could have suspected, that animal welfare is conflated with holistic conceptions of human health. Some theorists of animal welfare propose analyses of animal welfare that come close to biologically oriented analyses of human health.

Broom on welfare

Let me first introduce Broom's basic definition of welfare: 'The *welfare* of an individual is its state as regards its attempts to *cope* [my italics] with its environment ... State as regards its attempts to cope refers to how much has to be done to cope and how well and how badly coping attempts succeed' (1991, p. 4168).

> This definition of welfare has several implications 1. Welfare is a characteristic of an animal, not something that is given to it; 2. Welfare will vary from very poor to very good; 3. Welfare can be measured in a scientific way that is independent of moral considerations; 4. As explained above, measures of failure to cope and measures of how difficult it is for an animal to cope both give information about how poor the welfare is; 5. A knowledge of the preferences of an animal often gives valuable information about what conditions are likely to result in good welfare, but direct measurements of the state of the animal must also be used in attempts to assess welfare and improve it; and, 6. Animals may use a variety of methods when trying to cope. There are several consequences of failure to cope, so any one of a variety of measures can indicate that welfare is poor, and the fact that one measure, such as growth, is normal does not mean that welfare is good.
>
> (Broom, 1991, p. 4168. See also Broom, 1988, 1993a, 1996)

Coping in its turn is given the following characterization in Broom (1993a):

> Coping means having control of mental and bodily stability. Attempts to cope include the functioning of the body repair systems, immunological defences, and emergency physiological reactions as well as behavioural responses ... Coping may be difficult or attempts to cope may fail, with the result that there is *reduced biological fitness* [my italics]. In either of these two cases there may be suffering. Alternatively, an individual may be coping effectively with all aspects of life so that the terms happiness or contentment might be used in describing it.
>
> (Broom, 1993a, p. 16).

The mental and bodily stability that is the successful result of coping is thus related to the animal's biological fitness. Stability, in this conception, then means that the animal's state of stability is a sufficient condition (given standard circumstances) for its biological fitness. Fitness in its turn is characterized in the following way by Broom: 'the fitness measures such as survival, growth and reproduction have often been quoted, in this way, showing that welfare is good' (1988, p. 16). '[The animal] may fail to cope in that its fitness is reduced as evidenced by death, or failure to grow, or failure to reproduce' (1991, p. 4168). It is evident that Broom's notion of fitness here is not identical with that of Barnard and Hurst. Reproduction is only one of the criteria of fitness in Broom's theory. (Cf. however Broom and Johnson's, 1993, p. 66, formal definition of fitness which is made strictly in terms of evolutionary theory.)

Thus, there is a strong conceptual relation between Broom's notion of welfare and the notion of biological fitness. As we have seen, a similar relation holds in Christopher Boorse's theory of health: an organism is healthy if and only if all its organs contribute in a species typical way to its fitness, i.e. the survival of the individual or the species. (Boorse has a slightly more narrow conception of fitness than Broom.)

It would be too rash, however, to identify Broom's theory of welfare with Boorse's theory of health. This is so for two reasons. First, Broom has a slightly different vocabulary than Boorse, which may indicate conceptual differences. Second, Broom says more about welfare (see below in this chapter), which indicates a concept of welfare that is more complex than Boorse's notion of health.

But let me first go more deeply into the basic concept of welfare. Broom says: the animal's welfare is its state as regards its attempts to cope with its environment. This is slightly ambiguous. Is Broom talking about the animal's state as a *condition* for coping or is he talking about the animal's state as a *result* of its coping, or both? The first interpretation can be paraphrased as follows:

1. An animal's welfare is the state that enables it to cope (or disables it from coping; I follow Broom's idea of welfare as a dimension and not just an optimal state) with the environment in order to maintain biological fitness.

If this is the interpretation to be chosen it comes very close to Boorse's theory of health. (In correspondence, 2005, however, Broom rejects this interpretation.) The second interpretation can be paraphrased thus:

2. An animal's welfare is the more or less successful or unsuccessful *result* (or rather: the continuous *results*) of its coping attempts.

This interpretation is different from what Boorse had in mind. And it can be very saliently different if we include certain phenomena, such as subjective well-being, in the results. In earlier papers Broom did not explicitly mention subjective well-being. Later, however, feelings are described as important elements of coping. 'Feelings are a part of the biology of the individual that has evolved. They are used in order to maximize its fitness, often by helping it to cope with its environment ... Feelings are not a minor influence on coping systems, they are an important part of them' (Broom, 1998, p. 393). I will later discuss the role played by feelings in Broom's theory of welfare.

Analytically the two interpretations are quite different. However, empirically, it may be difficult to hold them separate. Consider the following schematic series. The mind and body of animal A cope successfully at time t0; as a result it is in state S at time t1; state S is a crucial condition for coping successfully at t1; as a result there is a state S1 at t2; S1 is a condition for coping successfully at t2, etc. In fact, it is problematic to talk in terms of a state of the body. There is no 'static' state in a biological organism. There is a continuous process. The process of coping at a certain moment is partly a result of the process of coping at a previous moment, and is itself a condition for the process of coping at a subsequent moment.

This gives us in fact a further (and perhaps the most plausible) way of analysing Broom's definition:

3. An animal's welfare is its process of coping with the environment.

This distinction between welfare as a process and as a state will become crucial in my further analysis of Broom's theory below.

Let me observe a further difference between Boorse's conception of health and Broom's conception of welfare. Boorse discusses coping in terms of physiological functioning only. He does not include the person's behaviour. Broom clearly includes the *behaviour* of the animal in the notion of coping. To have good welfare for Broom presupposes the emergence of coping behaviour. Good health for Boorse does not entail any particular behaviour on the part of the healthy person. (It is a different matter that the healthy person must behave in a particular way to *maintain* health.)

Let us now assume that the third interpretation of Broom's definition of welfare comes close to a fair interpretation. This entails that Broom's welfare has an agentive component. The behaviour of the dog, the cow or the pig, in terms of food-seeking, for instance, is part of its welfare. This reminds a philosopher of Aristotle's conception of welfare, viz. *eudaimonia,* which consists of the human being's virtuous activity. Aristotle goes completely to the behavioural end, however: physiological functioning is not part of *eudaimonia* (see Chapter 3).

The agentive or active component of Broom's welfare distinguishes it from many modern accounts of human quality of life. Quality of life is often interpreted as happiness/unhappiness (as an emotion or as a disposition for an emotion). Other interpretations include further components, in addition to happiness, such as external conditions and human relations, but rarely human activity as such.

Animal welfare and indicators of welfare. Definitional and other criteria in Broom's theory

Above I have given a provisional account of Broom's conception of welfare. In his treatments, however, he says much more about welfare and indicators of welfare. An essential task for Broom is to make welfare an operative concept; he claims that it is possible to measure welfare in a scientific way. For this purpose he needs to identify indicators or what he calls measures of welfare. It is of course crucial that these indicators or measures match the definition in order to be meaningful.

A first requirement is to become clear about the very notion of an indicator or a measure. First, we have to analyse the (logical or empirical) strength existing between the indicator and the thing indicated. Consider the following differences:

1. A is an indicator of B, means that when A is the case, then B must be the case. This may hold when B follows logically from A. We have a few such cases below.

2. A is an indicator of B, means that when A is the case, then, with some probability, B is the case. This may be when B is a typical (but not universal) effect of A.

Second, we may differentiate indicators depending on the type of connection that holds between the indicator and what is indicated. Consider:

3. A is an indicator of B, means that there is a logical (including conceptual) relation from A to B or from B to A.

4. A is an indicator of B, means that there is a causal relation from A to B or from B to A.

Broom and Kirkden (2004) present the following list of measures of welfare:

- Physiological indicators of pleasure.
- Behavioural indicators of pleasure.
- Extent to which strongly preferred behaviours can be shown.
- Variety of normal behaviours shown or suppressed.
- Extent to which physiological processes and anatomical development are possible.
- Extent of behavioural aversion shown.
- Physiological attempts to cope.
- Immunosuppression.
- Disease prevalence.
- Behavioural attempts to cope.
- Behaviour pathology.
- Brain changes, e.g. those indicating self-narcotization.
- Body damage prevalence.
- Reduced ability to grow or breed.
- Reduced life expectancy.

Some of these measures follow the definition closely. This holds for reduced life expectancy and reduced ability to grow and breed. The same holds for physiological and behavioural attempts to cope. More interesting and debatable are the following measures, which I will analyse in some detail.

Disease, injury and abnormal behaviour and their relation to welfare

Broom states explicitly that disease, injury and abnormal behaviour are indicators of poor welfare. I quote: 'The welfare of any diseased animal is less good than that of an animal which is not diseased' (Broom, 1996, p. 25). In an earlier paper (Broom, 1988, p. 13) Broom claims that the welfare of *most* diseased animals is poor.

In the context of discussing the role of suffering in welfare, Broom makes a comment on injuries. 'An individual might be injured without feeling pain because endogenous analgesic opioids, or indeed artificial analgesics, prevent the pain. However, if injury occurs the state of the animal is affected and welfare is poor' (1991, p. 4168).

Broom makes an explicit definition of abnormal behaviour and also discusses its relation to welfare:

> Abnormal behaviour is behaviour that differs in pattern, frequency or context from that which is shown by most members of the species in conditions that allow a full range of behaviour ... An abnormal behaviour might help an individual to cope, but it is still an indicator that the animal's welfare is poorer than that of another animal that does not have as much difficulty in coping.
> (Broom, 1991, p. 4171)

What is striking here is that Broom makes *universal* claims about the relations between disease, injury and abnormal behaviour on the one hand, and poor welfare, on the other. (The exception is the 1988 paper on disease.) The question, then, is whether such a universal relation is conceptual or empirical. An empirical universal claim is difficult to substantiate, and Broom does not attempt to do so. It is always possible, however, to make a conceptual claim of the kind: to have a disease *means* to have welfare lower than ideal. On the other hand, such a move can complicate the theory.

Let me here test the idea that a disease must for conceptual reasons reduce good welfare. This has implications for the concept of disease. Given the theory, everything that reduces good welfare must, per definition, compromise the individual's coping process, viz. its process of maintaining fitness. Is this true of all diseases? Again, we can *define* a disease to be such a process as actually (and not only potentially) compromises an individual's coping process. (This seemed to be Broom's position in conversation, January 2003.) Broom's explicit definition of disease is to be found in Broom and Kirkden (2004) (see above, Chapter 7). Then, the theory remains intact. On the other hand, if we follow common parlance and define disease ostensively by calling upon conventional lists of diseases, then we get into trouble. Some diseases, for instance diseases of the skin, are so mild that they never affect the individual. Early stages of diseases need not affect coping, and other diseases are aborted before they create trouble. *Mutatis mutandis*, the same reasoning applies to injuries, not all injuries affect their bearer's competence in coping.

In a late paper Broom and Kirkden (2004) seem to produce a more definite answer on this point. Here they explicitly state that health is a part of welfare. Whenever there is a reduction of health there is a reduction of welfare and health, as we have seen, is in its turn characterized as an animal's state as regards its attempts to cope with pathology. So, if the animal cannot eliminate its pathology, i.e. if it remains diseased for some time, then its welfare is reduced during this period.

Turn now to abnormal behaviour. Broom says, in the quotation above, that abnormal behaviour (or extent of suppression of normal behaviour) is an indicator that the individual's welfare is poor, even in the special case where the abnormal behaviour may serve the purpose of coping (Broom, 1991, p. 4171). Abnormal behaviour is explicitly called an indicator; Broom therefore hardly assumes that it belongs to the meaning of the term '(poor) welfare'.

The concept of normality assumed here is a statistical frequency concept. Abnormal behaviour is such behaviour as deviates from what most members of the species display. Thus, abnormality here neither refers (per definition) to fitness, nor is it a normative concept telling how an individual *should* behave. Why, then, would the abnormality in itself be such a strong indicator of poor welfare?

Many animals can exhibit peculiar behaviour. My own dog can occasionally, when he is extremely happy, keep running around in circles in front of me. This behaviour is unusual and according to the frequency definition of abnormal behaviour it is abnormal. But I would be surprised if this were taken as an indicator of poor welfare in the case of the dog. Moreover, Broom does not explicitly allow for *supernormal* behaviour in animals, for instance behaviour that displays extraordinary intelligence or extraordinary strength. Such behaviour would come out as abnormal in Broom's sense. However, we would hardly regard it as an indicator of poor welfare.

It may be noted that Christopher Boorse's concept of normality combines the idea of statistical normality with the idea of fitness: such organ function is normal as makes *at least* the species-typical contribution to *human survival and reproduction*. For him the supernormal functions thus come out as healthy. Perhaps this notion of normality can serve the purpose Broom is asking for. If this notion is chosen, then the idea that normality is an indicator of a high degree of welfare will be a logical consequence of the theory.

The relationship between suffering and welfare

It is a plausible idea that subjective well-being and suffering should be central in the definition of welfare. This is a central thought in the theory of human quality of life and it is also central in much animal science theory. Dawkins (1988, p. 209) states: 'To be concerned about animal welfare is to be concerned with the subjective feelings of animals, particularly the unpleasant subjective feelings of suffering and pain'. And Duncan (1993, 1996) notes that we speak of welfare only for organisms that are capable of experiencing subjective feelings, and he proposes that animal welfare should be defined only in terms of what the animal actually experiences. 'Neither health nor lack of stress nor fitness is necessary and/or sufficient to conclude that an animal has good welfare. Welfare is dependent on what animals feel' (Duncan, 1993, p. 12).

Broom does not (entirely) belong to this school of thought. Phenomenological well-being or suffering is not explicitly mentioned in his (basic) definition of welfare (1986). On the other hand, suffering has a prominent place in his writings and it is clearly encompassed in his conception of welfare (1998 and later writings). Broom (1998) demonstrates the significance of feelings in the animal's life, not just as *epiphenomena* of physiological processes but also as *essential* elements in the coping mechanism. This holds in particular for feelings such as pain, hunger and thirst. But almost all feelings (be they positive or negative) can play a positive coping role for the individual.

But although feelings can be parts of coping systems, they do not make up all of them (Broom, 1998). So if the concept of welfare is to be usable it must refer to all aspects of coping systems and not just to feelings. Let me point to a major passage where Broom discusses the limited role of suffering and feelings in general. Broom (1991) makes the following observation:

> Suffering and poor welfare often occur together but welfare is a somewhat wider term. Unpleasant subjective feelings will clearly affect the state of an individual as regards its attempts to cope with its environment. However, it could be that the state is affected without suffering occurring. Five examples of situations in which welfare can be poor in the absence of suffering are described briefly below.
>
> (Broom, 1991, p. 4168)

(As a comparison, Broom, 1993b denies that suffering need affect coping.)
    I here summarize Broom's five examples:

1. There may be injury without pain.
2. Suffering might be interrupted by sleep. The welfare is still poor during the sleep.
3. If there is impaired immune system functioning, but no suffering, then welfare is still poor.
4. Inability to reproduce and risk of premature death (without suffering) indicates poor welfare.
5. If there is little normal behaviour but no pain because of self-narcotization, then there is poor welfare. (Broom, 1991, pp. 4168–4169.)

Consider these arguments for *not identifying* welfare with subjective well-being/suffering. In general, Broom is here faithful to his basic definition of welfare in terms of coping for fitness. Argument 4 simply states a logical truth, given his definition of welfare. Statements 1 and 3 also follow, given the interpretation that poor health necessarily means poor welfare. Item 5 can, however, be criticized from the point of view given above. Abnormal behaviour need not compromise coping. But Broom does not introduce a new idea of welfare here. Argument 2 is special in that it points to the fact that subjective well-being and suffering can be interrupted by sleep without any basic biological change taking place.
    I would argue here that one may consistently have a feelings-based approach and allow for periods of sleep. In the theory of emotions it is recognized that we use locutions such as 'Liza was happy all week' and 'John has been depressed for several months now'. These locutions do not imply that Liza and John did not sleep at all during the periods covered. They entail that Liza and John when awake (most of the time) felt happiness and depression. What existed continuously, however, were *dispositions* for happiness and depression. Even during sleep Liza was disposed to happiness, meaning that if somebody (gently) woke her up she would again feel happiness. Thus, although a feeling is not present during a whole period, a disposition for that feeling is present all the time. This means that the *concept* of this particular feeling is relevant for the characterization of the whole period (see Nordenfelt, 1994).

Broom and Kirkden (2004) say:

> If the definition of welfare were limited to the feelings of the individual as has
> been proposed by Duncan & Petherick (1991), it would not be possible to refer
> to the welfare of a person or an individual of another species which had no
> feelings because it was asleep, or anaesthetized, or drugged, or suffering from a
> disease which affects awareness. A further problem, if only feelings were
> considered, is that a great deal of evidence about welfare – like the presence of
> neuromas, extreme physiological responses or various abnormalities of behavior,
> immunosuppression, disease, inability to grow and reproduce, or reduced life
> expectancy – would not be taken as poor welfare unless bad feelings could be
> demonstrated to be associated with them. Evidence about feelings must be
> considered for it is important in welfare assessment, but to neglect so many
> other measures is illogical and harmful to the assessment of welfare, and hence
> to attempts to improve welfare.
>
> (Broom and Kirkden, 2004, p. 339)

Here Broom and Kirkden do not recognize that the scope of the feeling argument can be made much broader. Most disease states are such that they tend
(in the long run) to create suffering, although they need not initially do so. Or,
most diseases are instances of a type whose majority of instances do (in the
long run) create suffering. In human affairs, such a tendency can be *the very
reason why we designate* a state of affairs a disease. Thus, the concept of suffering could (or as I would say: ought to) have a much more central position in
the theory of welfare than the one allotted by Broom. This is the point of the
reverse theory of illness that has been described elsewhere (Fulford, 1989;
Nordenfelt, 2001) and to which I will return in Part III.

The place of suffering in Broom's theory

As we have seen, Broom is eager not to identify poor welfare with suffering.
There can be poor welfare in terms of reduced coping without any suffering
occurring. On the other hand, Broom strongly emphasizes the significance of suffering when it occurs. The subjective feelings of an animal, according to Broom,
are an extremely important part of its welfare. 'Suffering should be recognized and
prevented wherever possible' (1996, p. 26). And, more importantly: 'If there is
suffering then welfare will *always* [my italics] be poor' (1993a, p. 22).

The order of deduction goes here from suffering to poor welfare. In his
arguments above Broom has discussed the order from poor welfare to suffering. But why is the former relation universal? Is it an empirical relation or is it
some kind of conceptual connection?

If we state that suffering is always an indicator of poor welfare, then we
must, following Broom's definition of welfare, claim that suffering is always an
indicator of poor coping. This, however, is hardly true and is indeed contra-
dicted by Broom himself. In his 1998 paper on 'Welfare, stress and the evolution of feelings' Broom goes into great detail in showing how almost all
feelings have evolved as essential elements in coping. Broom does not claim

that all instances of feelings are elements of good coping, but almost all *types* of feeling can be part of coping strategies. This holds not least for pain:

> The importance of feeling pain in promoting individual survival is considerable ... When pain is felt, the individual can take action to minimize tissue damage being caused, and the greater the feeling of pain, the faster the initial action is likely to be. Once pain has been felt in a recognizable situation, the possibility of learning to avoid any future pain, and hence damage, of the same kind is increased.
> (Broom, 1998, pp. 379–380)

The feeling of fear is a normal response to consciously recognized external sources of danger. This feeling can serve as an alarm for escaping behaviour but also result in reduced behavioural and physiological activity so as to render the animal inconspicuous to a predator. In both cases the result is functional. Indeed, malaise or the general feeling of illness can have adaptive effects:

> When the immunological and other defences of the body are having to work hard and consume a lot of energy, it is advantageous for the individual to rest and to be able to concentrate available energy on fighting the cause of the malaise. Even high fever would be adaptive if the net benefit from killing pathogens and suppressing activity outweighed the net cost of tissue damage.
> (Broom, 1998, p. 381)

This observation creates a tension in Broom's theory. When pain, even serious pain, contributes to a coping process, then it should be part of good welfare. However, this contradicts Broom's earlier statement that 'whenever there is pain there is always poor welfare'. Thus by explicitly introducing feelings into his notion of welfare he has not *ipso facto* arrived at the conclusion that all suffering entails reduced welfare. Intense suffering that is functional should be indicative of good welfare. If Broom's crucial statements on this point were to be harmonized then we would have to talk about two concepts of welfare, one that has to do with coping and one that is feeling-related. I doubt that Broom would like this conclusion. The general tendency in his writings is to stick to the basic coping conception of welfare.

Let me go more deeply into this problem through the following example. A horse that is basically strong and fit is put into a harsh environment. It must work hard to find enough straw for its subsistence and the climate is severe. The horse suffers frequently, because it is often hungry. Moreover, the horse often shivers from cold. But, *ex hypothesi*, the horse copes well. It struggles successfully to find food. And in spite of the climate and the cold, it never contracts any diseases. Thus, the horse copes successfully, according to the definition. But because of the environment the horse frequently suffers considerably. Should we then say that it has a high degree of welfare?

Broom might answer that this horse is an example of an animal that has *difficulty in coping*. There is low welfare, he says, not only when the animal fails to cope but also when it is having difficulty in coping (Broom and Kirkden, 2004, p. 338). It is unclear, however, what will count as difficulty. Can a negative feeling in itself be an indicator of difficulty? Broom and Kirkden (2004) are eager to emphasize that feelings are aspects of an individual's biology that must have evolved to help in survival. 'Aspects of suffering contribute significantly to how the individual tries to cope', they say (p. 339). There are interpretations of this statement that go in two different directions. One obvious interpretation is that suffering often prevents animals from coping well. They may become disabled because of the suffering and are therefore prevented from properly taking care of themselves. But another crucial interpretation is that *suffering*, in terms of for instance hunger, *may be a necessary part of successful coping*. The hunger alerts the animal to the fact that it needs food and makes it seek nourishment (cf. Simonsen, 1996: 'I am not convinced that a coping animal necessarily feels well').

In order to disentangle this problem I think we must distinguish clearly between coping as a *process* concept and coping as a *final state* concept. To *cope well* can mean different things given the two interpretations. The suffering horse can be coping well in the sense that it works efficiently towards reaching a final state of stability that will in the end (hopefully) be there when the harsh environment is improved. Thus, when coping is taken in its process sense then a suffering animal may indeed be coping well. The hunger and the pain of the horse (if the suffering is not too severe) may indeed provoke the horse and speed it up in its endeavour. However, if we consider a particular point in time and 'freeze' the process the situation may be interpreted differently. Assume that we are to judge a particular moment when the horse feels a lot of pain and is very hungry. We may then ask whether we consider this moment to be one where the horse has coped well in the sense of having achieved stability. The answer is no. The horse does not have stability at this moment.

The question now is how we should assess the two types of coping in terms of welfare. I have tentatively suggested that coping in the process sense (i.e. possibly entailing some suffering) could constitute good welfare. However, if we accept this then, given Broom's theory of welfare, we may end up in a contradiction. An animal could have both good and bad welfare at the same time. It has good welfare at time t because at t it is involved in an efficient coping process. It has bad welfare at time t because at t it is in a state when it has not achieved stability.

In my own theory of holistic health, to be discussed in Part III, I argue that all significant suffering will negatively affect a person's ability to perform *intentional actions*. Therefore significant suffering will always entail reduced health according to my definition. But Broom's case can hardly be helped by this idea. His basic theory is different. He talks in terms of coping, including physiological functioning and either automatic behaviour or behaviour which involves cognition. It is not true in general, as Broom himself concedes, that an animal in significant pain would be deprived of coping strategies. In fact, the animal could in some instances be alerted and more effective. (See also Duncan, 1996, pp. 31–34, where he discusses both positive and negative instances of stress.)

Broom on health and welfare: a comparison

I have earlier presented Broom's characterization of the notion of animal health and I have compared it with the definition of human health proposed by Christopher Boorse. I will now raise the question how health and welfare can be distinguished in Broom's (as well as in, for instance, McGlone's and Barnard and Hurst's) systems. Is there a risk of total identity? If that were the case these notions could hardly be employed, as Broom claims, also in the human context. Welfare or quality of life has normally a meaning that lies far above health in the human context. Thus there is a case for looking into the real differences between animal health and animal welfare. I will approach this problem by further analysing the notion of coping as used by Broom and others, and particularly look into Broom's notion of degrees of coping and Boorse's notion of contributions to survival and reproduction.

There is one important observation to be made here. Health for Boorse is a threshold notion. As soon as one has reached full health, health can no longer be improved. When all organs (including mental faculties) in an individual contribute, *at least* in a species-typical way, to survival and reproduction, then the individual is completely healthy. For the case where some or even all organs contribute *better* than this, Boorse offers no description in health terms. Welfare for Broom, on the other hand, is not a threshold concept of the same kind. Animals can improve their welfare much beyond a statistical average. The supernormal animal that copes extremely well can have a high degree of welfare. (This interpretation was made clear by Broom in an interview with him in January 2003.) Broom's theory does not, however, indicate that there is welfare beyond coping.

Could the following then be the most fruitful interpretation of Broom and the coping theorists? Health constitutes the lower part of welfare. Poor health is *ipso facto* poor welfare. Good welfare is something over and above complete health. Before resting with such a conclusion, however, I will scrutinize the notion of coping a little more. First, coping for the animal theorists does not only involve physiology (or psychology) as with Boorse. Coping involves also behavioural strategies. Thus, poor coping is not just poor health à la Boorse. The badly coping animal also *performs* badly in relation to survival, growth and reproduction. (The badly coping animal normally also feels bad, but this hardly constitutes a difference in relation to Boorse's analysis. Boorse does not deny that a person with a disease is often also feeling ill.)

The difference between Boorse's and Broom's conceptions of health, however, becomes weaker as we focus on mental health. The mental faculties express themselves in terms of behaviour. The mentally ill person often behaves in ways that are not fruitful for his or her survival or reproduction and these behaviours are often included in the very definitions of the illness. Thus, several mental illness concepts entail that there is a reduction in the individual's behaviour in relation to the biological goals of survival and reproduction.

Consider now the upper part of the scale, what is beyond Boorse's complete health, viz. the dimension from moderate welfare to a high degree of welfare with Broom. This is the situation where the animal is coping exceedingly well. The physiology is better than average, the behaviour is sharply focused on the ultimate goals of survival, growth and reproduction and the animal is extremely efficient in attaining these goals. Feelings can play a part in this coping (according to Broom). Both positive feelings, such as joy, and negative feelings, such as pain, can contribute both positively and negatively to the coping process.

Can this, however, be a complete analysis of the upper welfare scale? Let me contrast two kinds of situations. (I am here adding to the argument where I discussed the place of feelings in welfare.) One horse (A) lives in abundance, in extremely advantageous circumstances when it comes to nutrition, exercise and shelter. Another horse (B) lives in poor circumstances in the same respects. A is living a varied life, can choose between living indoors and out-doors, is often taken out for a ride and is provided with straw of excellent quality. B is continuously outdoors even in extremely harsh conditions; B is never taken out for a ride and must move about for himself, and the food B gets is very sparse. A is coping well with his environment without much effort. But, in my example, B also copes well. B is a strong horse that can sustain the harsh climate (although he frequently shivers) and B gets his exercise on his own and can manage with the inadequate food. B never succumbs to disease.

Both A and B can be said to be coping exceedingly well in the process sense. But do they have the same welfare? A is never suffering but B is often suffering. Suffering is a part of B's excellent coping process. But does B have the same welfare as A? Again, we ought to rely on our distinction introduced earlier. B is not coping as well as A in the final state sense of coping. A is in a stable state at all moments during our inspection. However, at almost what-ever moment we select, B is not in a stable state. Thus, in this sense A has good welfare whereas B does not.

Clearly, the interpretation of coping is crucial. Should the process inter-pretation of coping have a position equal to the interpretation of coping as a final state in the last analysis of welfare? If we stick to Broom's dictum that suf-fering always entails some reduction of welfare, then we must have the final state interpretation of coping in mind.

This whole discussion must in the end be related to the ethical concerns with regard to the welfare of animals. Much of the ethical discussion in animal welfare concerns external factors such as housing, nutrition and treatment in general. We say that these can be detrimental to the welfare of the individual animal. What welfare are we then talking about? Are we concerned about the animals that are not coping well, and then in what sense? Are we concerned about all suffering animals? Or are we only concerned about those suffering animals that are not coping perfectly well in the process sense of coping?

I will now raise a different argument regarding the upper part of the scale of welfare. Consider an animal that is perfectly at ease. Look at the young dog that is playing happily with its owner. This dog shows *every* sign of harmony and happiness. We assume that the dog has itself started the game; it has not been provoked by its owner, so there is no question of regarding the play as coping with a provocation. Certainly, the dog can be characterized as coping well with the environment. Its physiology is perfect and it exhibits the behaviour necessary for surviving *when this is called for*. But my point here is that animals that live in advantageous circumstances do not necessarily have to cope, with regard to behaviour, at all times. This is perhaps much more clearly seen with humans. There are times when we as humans do not have to cope at all (from a behavioural point of view, that is; our physiology is of course always in a coping process). We just live, and we sometimes live in pleasure and harmony. We feel well. I think that this is how we can also describe the playing dog.

The playing behaviour of the dog is therefore a kind of surplus behaviour; it lies over and above coping. This behaviour is an expression of happiness. I suggest that this must be included in an upper level of a welfare scale. I cannot see, however, how this state can be accommodated in a scale that is solely based on coping. The happiness involved in playing need not be part of the dog's coping process. And the happiness is something over and above a state of sheer stability.

I can envisage some counter-arguments here. Consider the following. The playing behaviour is in fact a way of coping with the environment. When the dog is at ease it has surplus energy that has to be released in some way. If the dog were prevented from playing it would in the end be frustrated and suffering and its coping would be reduced. Thus the playing must be regarded as a way of coping with the environment, at the upper end of the scale of welfare. (For a detailed contemporary discussion of play, see Spinka *et al.*, 2001.)

I agree that there are such instances of playing, i.e. where the playing behaviour has a function in a coping process. From this, however, it is a long way to saying that *all* playing behaviour has a coping function. Assume that our happy dog has a lot of opportunities for playing and that it frequently uses these opportunities. I assume also that at the moment I am focusing on the dog there is no *need* for playing in order to prevent future suffering.

Consider a further counter-argument. Although not all playing behaviour is strictly necessary for good coping, it has a role in strengthening the animal and in building up a better coping ability for the future. Thus the playing is anyhow in the end part of a more long-term coping process. I agree that there is a plausible speculation behind a statement of this kind. Similar statements are being used in popular health-promoting slogans: if you are a happy person and if you do what pleases you, then you will also become a healthy person.

But is it reasonable, given our present state of knowledge, to promote such a slogan as a general law? Is it reasonable to reduce all activity of animals and humans to a matter of coping for survival, growth and reproduction? Are all those activities that I as a human being like to do, and which raise my quality of life, such as reading books, travelling abroad and watching birds, also conducive to my basic coping? I can grant that some of them may to a slight extent indirectly be so. But what reasons do we have for believing that *all* liked activities are in the end coping-promotive? Indeed, people may consciously choose to act in ways that are detrimental to health and coping. A smoker who chooses to smoke to raise his or her quality of life is often well aware that the smoking may have negative health consequences in the end.

My conclusion therefore is that there is subjective well-being (and suffering) which is not a part of a coping process; therefore a theory of coping cannot be a complete theory of human or animal welfare, in which we believe that subjective well-being and suffering play a crucial role.

10 Theories of Welfare, Ethics and Values

The biologically oriented theories of animal welfare have highlighted a crucial problem in the theory of welfare. The proponents of such theories, including Barnard and Hurst and Broom, claim that the concept of welfare can be given a purely scientific characterization without any evaluative component. Other theorists, in particular the ones on the well-being side, deny that such definitions are possible. This controversy has a counterpart in the philosophy of human health and welfare. I will here consider the arguments *pro et con* and attempt to take a stand on the issue.

Donald Broom has explicitly taken a stand in this debate in favour of the scientist position:

> Welfare varies on a continuum from very good to very poor, and it can be assessed precisely. The assessment of welfare can be carried out in a scientific way without the involvement of moral considerations. The question which is asked after the measurement is made is 'how poor must the welfare be before people consider it to be intolerable'. A moral decision must be taken here and different people will draw the line, marking what is unacceptable, at different levels in the welfare continuum. The moral decision depends upon the availability of evidence about welfare, but the process of deciding about morality and the process of assessing welfare are quite separate.
>
> (Broom, 1988, p. 6)

In an interesting article Tannenbaum (1991) denies the validity of such a position. He argues that 'the pure science model fundamentally misconstrues the nature of animal welfare' (p. 1360). Tannenbaum denies the possibility of studying animal welfare scientifically without considering or taking any positions with respect to ethical issues. He exemplifies his thesis in many ways. I give here just the following example:

> Welfare investigators are not interested only in scientifically determining various levels of welfare. If this were really their interest they might, in fact, study

conditions that had no possibility of being adopted. Investigators would not propose such studies because doing so would be practically pointless, and animal welfare scientists make the *value judgment* that it is not appropriate to conduct pointless investigations.

(Tannenbaum, 1991, p. 1365)

Tannenbaum then goes on to analyse the following position. Scientists might agree that the decision of an acceptable level of welfare is an ethical decision. However, they may still insist that once this decision is made one can carry out the study without reference to values. Tannenbaum objects also to this position. 'The very concept of animal welfare – what ordinary people and scientists mean by the term welfare – includes an ethical component' (1991, p. 1366). Tannenbaum observes that welfare refers to what is good for somebody, and goodness is the most central evaluative concept.

David Fraser (for instance in 1995 and 1997) has developed this point considerably. He says (1995, p. 104) that an attribute is related to animal welfare if it is somehow good or bad for the animal. Thus the concept of animal welfare is unlike the many scientific concepts (such as temperature and viscosity) that can be quantified without necessarily invoking any sense of better or worse. Fraser devises a typology of concepts, type 1, type 2 and type 3 concepts to explicate the difference between scientific and evaluative concepts. Welfare is a type 3 concept that presupposes values. As a consequence, according to Fraser, we can never simply measure animal welfare. We can only directly measure such variables (of type 1 and type 2) as are conducive to states of affairs that we evaluate to be *good* for the animals. (I will return to this in a special section on Fraser's theory of animal welfare, Chapter 17.)

I am basically in favour of Tannenbaum's and Fraser's position but I wish to make some, hopefully, clarifying distinctions. There is an aspect of the pure science view that is reasonable, and not recognized by Tannenbaum (nor discussed by Fraser).

First, I think it is of importance to distinguish between an ethical decision and a value judgement. The philosophy of values covers a much wider ground than ethics. There are many evaluations outside ethics. There are aesthetic, religious and intellectual evaluations that have little or nothing to do with ethics. To say that something is beautiful, sublime or well-organized is to evaluate, but it is not to evaluate in an ethical sense. Moreover, there are certain evaluations that are highly relevant to ethics without being *ipso facto* ethical. Central among such evaluations are the welfare ones. One can say that a certain state of affairs has a high welfare value without making an ethical judgement. The ethical judgement comes in when one decides that it is morally good to aim at this state of affairs. But it is possible to have a different position, for instance one following the German philosopher Immanuel Kant, according to which aiming at welfare is not, as such, ethically good.

I would therefore qualify Tannenbaum's position in saying that a welfare evaluation can be made without any ethical decision having been taken. However, this need not differentiate us much in relation to the pure scientists.

These people, including Broom, may still contend that determining that state X is a state of welfare is a purely scientific matter, not involving any evaluations, be they ethical or otherwise.

To take a stand here I wish to distinguish between primary and secondary assessments of welfare. A primary assessment must involve an evaluation. A secondary assessment relies on a primary assessment but does not itself express an evaluation. Let me illustrate. Assume that a theorist has decided that coping for fitness (in the sense of survival, growth and reproduction) is what should be meant by animal welfare. This decision (or this choice of definition) involves an evaluation. It entails saying that coping for fitness is a *good* state to be in for an animal: it is the faring well of this animal. (On this point I entirely agree with Tannenbaum.) The scientist who makes this definition for his or her further scientific purposes has made a primary assessment and is committed to an evaluation.

However, once this primary assessment has been performed there are many things to follow of a non-evaluative, i.e. 'purely scientific', nature. One can ask many causal questions concerning what environments, or what physiological states, contribute to fitness. In a secondary way the scientist can thereby assess if states X and Y are states of welfare or on what level on the welfare scale they are. This scientist need not be committed to the primary evaluation. He or she need not even agree with this evaluation.

It should, however, be noted that it is difficult to avoid evaluations even on the secondary levels. This has to do with the fact that it is many times more difficult to answer precisely the causal, scientific questions. And in such a case the scientist may hypothesize that a state X is conducive to welfare for other, non-scientific, reasons. The following is an example. A certain condition has long been included in a conventional list of diseases. It is often assumed (without proper investigation) that all diseases compromise welfare (i.e. compromise fitness) (see my discussion below). The determination of this 'disease' as a state of illfare – I coined the term 'illfare' in 1994 (see Part III of this book) – is then not scientifically made.

Thus, although it has been disputed by some animal science theorists (e.g. Broom, 1991, and later Barnard and Hurst, 1996), I will argue that 'welfare' is an evaluative term. To fare well for an animal is to live a good life. This is the obvious lexical meaning of the words. When the theorists in question deny that this is so, it is a mystery why they choose the term 'welfare' and not 'coping for survival or reproduction' or one of the other notions that appear in the *analysanda* of their investigations. To designate coping for survival as welfare is to say – and I cannot find any alternative to this – that coping for survival is something *good*, something that we should enhance. And then one has admitted that using the term 'welfare' commits one to an evaluation. (For similar arguments see Tannenbaum, 1991; Fraser, 1995.)

We seem therefore to have the same basic investigation to make as I have sketched for the human case above. Do we think that there are absolute values? Is there an animal *eudaimonia*? Given Aristotle's general philosophy of nature there should be a perfect life for the animal which however would probably not be called *eudaimonia*, the term for the specifically human way of

faring well. Everything in nature has its place; there is a naturally given *telos*
(goal) for every object, including all animals. To be in the due process to
reach and maintain this *telos* is to fare well. Our problem is only to determine
what the *telos* of a particular animal is. This problem may turn out to be
formidable.

Certain biologists have been attracted by this view, however. The idea
then would be to assume that the biological *telos* is inscribed in the genetic
code of the animal in question (consider Rollin, 1992, and later, e.g. 1996).
By inspecting the lives of individual animals, for instance dogs, we might be
able to deduce what has been inscribed in the code, and that could tell us what
the good life for the dog is. The physiological mechanisms, including states of
homeostasis, show where to find the biological *telos* on the physiological level.
The behaviour of the dog, when it is unrestricted and in a 'natural' environment, will show us its biological *telos* on the ethological level, and so on.

The idea of a *telos* of animals, and the associated idea of natural behaviour, is intriguing and worthy of an analysis in the welfare context. It is particularly forceful when it is connected with certain procedures for determining the
telos. (I am thinking of signs such as pain or frustration to find out when an
animal's *telos* has been tampered with.) I will return to the general idea in
Chapters 15 and 16. There are, however, major difficulties with the notion of
telos. First, we have to ascertain whether we are talking about *telos* as a
concept pertaining to species or as a concept pertaining to individuals. Assume
that we use as a reference point the *telos* of the species of dogs in assessing
the welfare of a particular dog. We then face the problem of determining the
telos of dogs in such detail that it will be relevant to the individual. There is no
single nature of a biological species. All species have varying subspecies and
in the end varying individuals. The physiology of the individuals and, especially,
their behaviour may differ considerably. There is no neutral reference point.

One may perhaps lean on frequency estimations. However, doing that is
moving away from the spirit of the Aristotelian notion of an inborn *telos*.
Moreover, it is perfectly possible that the majority of a specific population
exhibits something that is far from what we would otherwise be inclined to call
welfare. The species may be a dying species. It may be a species that is coming
to a close, because it is no longer fit for the modern civilized world. As we have
seen, this is a tough point against any evolutionary platform for determining
welfare. Also a dying species has been 'designed' by evolution. Its present life,
however, can hardly be the criterion for anything called the welfare of this
species.

At most we can find a *telos* for each individual of a species given a fixed
environment. But then we can wonder whether we need the notion of a *telos*
for our purpose of determining the welfare of the individual. Is the ultimate
question for welfare whether the individual lives up to its more or less abstract
telos that we have tried to deduce from its previous behaviour? Are we not
more interested in the question whether the individual *likes* its present life? We
may perhaps answer, and I think Rollin would do so, that the 'liking' is a criterion of the *telos*, but then we have only complicated our reasoning with a
new abstract concept.

11

Theories of Welfare in Terms of Subjective Well-being

Feeling Theories of Welfare

Although Broom to some extent incorporates feelings in his theory of welfare, his basic notion is, as we have seen, the notion of coping. And if the basic approach refers to coping then feelings will come in only insofar as they contribute to or reduce coping.

Therefore it is a radically different approach to say that welfare basically has to do with the feelings of animals, whether they feel well or whether they are suffering. One of the strongest proponents of such a view is Marian Stamp Dawkins in, for instance, *From an Animal's Point of View: Motivation, Fitness and Animal Welfare* (1990):

> Let us not mince words: animal welfare involves the subjective feelings of animals. The growing concern for animals in laboratories, farms and zoos is not just concern about their physical health, important though that is. Nor is it just to ensure that animals function properly, like well-maintained machines, desirable though that may be. Rather, it is a concern that some of the ways in which humans treat other animals cause mental suffering and that these animals may experience 'pain', 'boredom', 'frustration', 'hunger' and other unpleasant states perhaps not totally unlike those we experience.
>
> (Dawkins, 1990, p. 1)

What we need now, she says, is to come to grips with the wide range of unpleasant emotional states we call suffering and the more pleasant ones we call contentment or well-being. The reason is that almost all arguments about the treatment of animals focus on the issue of suffering and the experience of pain (Dawkins, 1990, p. 1). Dawkins also mentions the relation between this idea of welfare and the animal rights movement. The idea that animals are morally the equals of humans and have rights comparable to those of humans is based on a belief in an equal capacity to suffer as well as on a belief that both have other similar attributes that influence subjective experience.

The most formal characterization of suffering that Dawkins presents is the following: 'Suffering occurs when unpleasant subjective feelings are acute or continue for a long time *because the animal is unable to carry out the actions that would normally reduce risks to life and reproduction in those circumstances*' (1990, p. 2). (As others have pointed out, this is quite narrow a use of the term 'suffering'.)

Dawkins's approach is shared by Ian Duncan, who notes that we speak of welfare only for organisms that are capable of experiencing subjective feelings: 'Thus, neither health nor lack of stress nor fitness is necessary and/or sufficient to conclude that an animal has good welfare. Welfare is dependent on what animals feel' (1993, p. 12). Duncan explains his position by arguing that welfare completely concerns conscious states, i.e. is the absence of states of suffering and the presence of states of pleasure. It therefore concerns the satisfaction of wants and desires rather than the satisfaction of needs (Duncan, 1996, p. 31).

A further proponent of a feelings-based theory is H.B. Simonsen, who proposes the following definition of animal welfare:

> Animal welfare consists of the animal's positive and negative experiences. Important negative experiences are pain and frustration and important positive experiences are expressed in play, performance of appetitive behaviour and consummatory acts. Assessment of animal welfare must be based on scientific knowledge and practical experience related to behaviour, health and physiology.
>
> (Simonsen, 1996, p. 92)

Other theorists agree that subjective well-being and suffering are the defining traits of welfare, but they find such states difficult to assess scientifically, whereas measures based on biological functioning provide an adequate and convenient means of obtaining relevant information. Gonyou (1993, p. 43) proposes: 'Although the animal's perception of its condition must serve as the basis for well-being, research in this area is only beginning. At the present time much can be accomplished by using more traditional approaches involving behavioural, physiological and psychological studies'. In contrast to this view, McGlone says:

> Recently, scientists have suggested that if an animal perceives that it feels poorly (as measured primarily by behaviour) then the animal is said to be in a poor state of welfare. I dismiss this view as simplistic and inappropriate. I suggest that an animal is in a poor state of welfare *only* when the physiological systems are disturbed to the point that survival or reproduction are impaired.
>
> (McGlone, 1993, p. 28)

This scepticism with regard to animal feelings and our methods for discovering feelings is even greater in the texts by Barnard and Hurst:

> At least three tenable possibilities therefore exist: a) other species can experience suffering in a form homologous with our own, b) other species experience

negative subjective states that are wholly different from our own but might lead to taxon-specific forms of experience that are analogous to suffering, and c) other species do not experience negative subjective states at all and cannot be thought of as having a suffering-like state in the sense of a) or b) … Anthropomorphic uses of the concept of suffering in welfare studies may therefore be dangerously inappropriate in particular cases.

(Barnard and Hurst, 1996, p. 410)

And as we have seen above, Barnard and Hurst identify welfare with development according to the evolutionary design.

The Problem of Animal Minds

We have seen that there is a controversy between leading theorists in animal science with regard to the appropriateness of using concepts referring to subjective states in animals. Some people, including Dawkins and Duncan, are confident that we can attribute suffering to animals and that suffering can be measured for scientific purposes. Others, like McGlone (1993), Barnard and Hurst (1996) and Bermond (1997) (to be discussed in Chapter 12), deny or at least doubt this possibility. According to them there is no 'scientific' argument for stating that animals can suffer in anything like the sense that we ordinarily believe. I will now take a closer look at this controversy. Consider first some significant statements by the participants in the discussion.

Peter Singer makes a blunt statement: 'Non-human animals can suffer. To deny this, one must now refute not just the common sense of dog owners but the increasing body of empirical evidence, both physiological and behavioural' (Singer, 1990, p. 9).

A.F. Fraser and D.M. Broom provide some pragmatic considerations:

To have a realistic perspective, no study of animal suffering should neglect that this is the central focus of clinical veterinary medicine. In its common use, the term 'suffering' covers the overtly manifested features that coexist with all clinical states that are substantial … Suffering can be regarded as the affective component of any significant disturbance of or insult to the subject's physiology or sentience. Most signs of suffering occur in behavioral syndromes that have long been recognized in veterinary medicine. Many involve pain, which often generates suffering. Pain is sometimes revealed in negative behavioral reactions, for example, when the seat of pain can be palpated. Fear, as an additional factor in the syndrome, may either subdue or exaggerate any expression of pain in animals. Shock can also mask pain. Behaviors that are well-recognized signs of pain include flinching, restlessness, flailing or rigid limbs, writhing or self-directed bites, and abnormal vocalization. Other signs of painful suffering include panic biting, suddenly altered appetite, inactivity, 'tucked up' posture or postural changes and modified motion.

(Fraser and Broom, 1997, p. 21)

G.M. Burghardt summarizes:

Both Dawkins and Singer advocate *critical anthropomorphism* … the explicit use of our own experiences and feelings, along with our knowledge of human

and nonhuman animals, empirical data, and the natural history of the species in
question, in approximating what other species experience and in generating
predictions.

(Burghardt, 1990, p. 11)

However, some animal scientists criticize this analogical reasoning, claiming
that it is not scientifically sustainable. We cannot 'prove' that the animals really
'feel' these sensations or emotions. We cannot 'prove' that they feel anything
at all. Thus, we should resist talking about suffering at all in scientific contexts.
And some of them conclude that we cannot equate animal welfare with animal
well-being in the sense of positive subjective experiences.

There are many problematic aspects of this attitude. I will list the major
ones here. (For an extremely interesting and well-written exposition of such
arguments and for an analysis of the psychology of scientific non-believers, see
Rollin, 1990b.)

1. We 'prove' nothing in empirical science. We make observations based on
hypotheses. These observations already entail some theoretical presuppositions, for instance a presupposition about minds.
2. A refutation of an argument from analogy would involve also a refutation
of the ascription of feelings to our fellow human beings.
3. The mental language is not a language of 'private' experiences. (This is a
position argued most forcefully by Ludwig Wittgenstein, 1953.)

I will discuss these and some further arguments in the following chapter on
animal minds.

12 On Animal Minds: a Digression

Some Classical Opinions

The discussion as to whether non-human animals have minds like humans has been going on for a long time. Aristotle did not hesitate. He discerned a great number of mental faculties in animals.

> For even the other animals mostly possess traces of the characteristics to do with the soul, such as present differences more obviously in the humans ... For some characters differ by the more-or-less compared with man, as does man compared with a majority of animals (for certain characters of this kind are present to a greater degree in man, certain others to a greater degree in animals).
>
> *(Historia Animalium* 588 a18–b3; for a modern edition, see Aristotle, 1991)

Aristotle distinguished between three kinds of minds (or souls, which is the common translation from the Greek). The vegetative mind exists already in plants, the perceiving mind is common to man and animals and the rational mind is unique to humans.

Descartes, on the other hand, denied the existence of an animal mind. As is well known, Descartes, in contrast to Aristotle, also drew a sharp dividing line between matter and mind, where mind is understood as a thinking substance, *res cogitans*. Animals and plants are mindless automata, says Descartes. The argument on which he relies is that only humans have a language by which they can express their thoughts and do so in original ways. Interestingly enough, Descartes emphasizes that all humans, even the most intellectually handicapped ones, have this capacity to some extent, whereas no animals, not even the cleverest among primates, have it at all (*A Discourse on Method*, part 5; for a modern edition, see Descartes, 1992).

It is worth noting that the empiricist and sceptic David Hume did not for a second doubt the existence of consciousness and mentation in animals: 'No truth appears to me more evident, than that beasts are endowed with thought and reason as well as men. The arguments are in this case so obvious, that they never escape the most stupid and ignorant' (Hume, 1961, p. 176).

Hume here certainly follows a common-sense idea that has hardly ever been much affected by science. It was also reaffirmed by Darwin and some of his successors. In *The Descent of Man* (1871), Darwin affirmed that there is no fundamental difference between man and higher animals in their mental faculties. 'The lower animals, like man, manifestly feel pleasure and pain, happiness and misery' (p. 148). In fact he attributes almost the full range of human subjective experience to animals. In *The Expression of Emotion in Man and Animals* (1872), Darwin makes a detailed investigation of the various bodily means by which animals and humans express their emotions. He gives numerous examples in particular from apes and dogs. In the case of the sensation of pain he says the following: 'An agony of pain is expressed by dogs in nearly the same way as by many other animals, namely by howling, writhing, and contortions of the whole body' (p. 122). About fear and terror he says: 'A dog under extreme terror will throw himself down, howl, and void his excretions' (p. 122). 'Even a very slight degree of fear is invariably shown by the tail being tucked in between the legs' (p. 122). And finally Darwin's comments on love and affection: 'With the lower animals we see the same principle of pleasure derived from contact in association with love. Dogs and cats manifestly take pleasure in rubbing against their masters and mistresses, and in being rubbed or patted by them' (p. 123).

Darwin's pupil George Romanes (1884) explores this territory further when he investigates the cognitive dimensions of animal experience, including the problem-solving abilities of earthworms.

Some Semantic Observations

Before introducing some elements of the modern discussion on animal minds let me comment on the notion of mind. It is evident that much of the discussion about the existence of minds, be they human or animal, has been plagued by semantic confusion. What and how much does mind contain? Descartes, in his labelling of mind as *res cogitans*, mainly focused on human thinking, i.e. the cognitive and also conscious aspects of mind. In certain modern discussions, in particular in cognitive science, mind and the mental is something quite exclusive, comprising only self-awareness and self-recognition. The best example of such an abstract discussion is perhaps the one introduced by Heyes (1998), where the issue is formulated in terms such as: do non-human primates have a *theory* of mind [my italics]? A commentator wondered with good reason how many humans have a *theory* of mind.

For Aristotle, and for many present-day theorists, the mind or the soul is something much more inclusive. Plants have a soul, although only a vegetative one, says Aristotle. This means that Aristotle's idea of mind/soul, as well as the theories of mind of many contemporary British philosophical psychologists (such as Ryle, 1949), does not presuppose *consciousness*. That a person has a mental property, such as being in love, need not entail that this person at every moment is conscious of anything. Indeed, Ryle and his followers would say that mental words normally do not refer to conscious states at all (see later in this section).

For many animal welfare theorists the whole area of animal cognition is fairly irrelevant. What is important for them is that animals can *suffer*, that they can feel pain, fright, depression and other negative sentiments, but also that they are capable of having corresponding positive feelings such as pleasure, ease and happiness. Welfare science thus introduces and emphasizes other areas of the mind than the cognitive ones. These areas are sensations, emotions and moods. (The grounds for distinguishing between these categories are given in Part III.) The welfare theorists normally understand these mental states as conscious states. The consciousness of pain and other suffering certainly is crucial for the whole welfare ideology.

An explicit and interesting statement to the contrary, however, is made by de Jonge (1997): 'positive or negative subjective emotions may exist without the organism being conscious of such emotions' (p. 104). Even suffering may take place without the organism being conscious of it. He explains: in so-called post-traumatic stress syndromes 'people may often show emotional behaviour (irritability, sleep-disturbances, depression [etc.]) without being conscious of the underlying emotions'. This explanation calls for a further distinction between two kinds of consciousness. The first kind of consciousness presupposes only that there is a feeling of some kind, either of a sensational kind, where the subject feels pain, feels cold or feels uncomfortable, or of a perceptual or emotional kind, the subject is aware of a danger or feels depressed. These feelings may be pleasant or unpleasant and, for instance, suffering can occur. De Jonge's case would then satisfy the criteria for this kind of consciousness.

The other kind of consciousness is of a second order. According to this idea a person can be *conscious of his or her feelings*. Presumably this means that the person is cognitively aware of the feelings. He or she is reflecting on the feelings. This kind of second-order consciousness is probably often absent in the case of most animals but also often in the case of humans. A human being can be suffering without being capable of consciously analysing or understanding what the suffering consists of.

Let me summarize. If mind is understood exclusively as conscious abstract thinking then it is unlikely that, besides humans, more than a few primates can compete. If, on the other hand, it suffices that an organism has (or can have) one or more of the following mental properties to some degree: sensation, perception, emotion, mood, inclination, will, thinking or believing, then the chance of including most kinds of animals with the possible exception of the unicellular ones is quite high.

The Development during the 20th Century

It is remarkable therefore that by 1930 there were few animal scientists speaking about animal minds and admitting that animals can think and have feelings. A major explanation of this is the emergence of a new scientific paradigm, positivism, which had devastating consequences for the whole field of psychology. (See, for instance, Ayer, 1978.) The principle of verification, included in this paradigm, requires that every scientific datum should be based on observations. Minds cannot be observed. Only objects and behaviour can be observed. Thus traditional psychological concepts, such as cognition, emotion and will, should be redefined in behavioural terms or be totally eliminated. This was the behaviouristic programme, introduced by J.B. Watson (1925).

Parallel to the behaviouristic programme emerged the ethological one in animal science. The main proponents of this programme were the two animal scientists Konrad Lorenz and Niko Tinbergen. An important forerunner was Julian Huxley. The ethologists were different from the behaviourists in focusing on the instincts of animals instead of their learning to behave. The ethologists, however, adopted a position similar to Watson's and the other behaviourists' in another crucial respect. They thought that the study of the subjective experience of animals was not essential for the explanation and understanding of animal behaviour. Moreover, they were eager to make the study of animals objectivistic. Thus the mind of animals disappeared from scientific research also in their case.

This did not entail, however, that all ethologists (in contradistinction to most behaviourists) denied the existence of animal subjective experience. Huxley has some famous passages (for instance, 1914) where he emphasizes how we can predict the emotions of birds much better than we can possibly deduce anything about their nervous processes. Moreover, Lorenz (1937) clearly ascribed subjective states to animals. He also believed that the subjective feelings constitute the immediate goal of appetitive behaviour. Lorenz furthermore expressed the hope that subjective phenomena could be susceptible of causal, scientific, explanation. (For a brief history of the ethological movement, see Burkhardt, 1997.)

The general scientific trend, however, was clear. The positivist (or objectivist) paradigm in principle struck human psychology as hard as it struck animal psychology. From a scientific point of view, positivism says, the only things that exist are observable bodies and their observable movements or behaviour. If we keep talking about cognitions and emotions in humans, as well as in non-human animals, we must mean certain sets of neurological events or certain sets of behaviour pertaining to these living beings. The consequences in practice for the ideas about minds in humans and animals, however, were different. In the case of animals, the scientific paradigm could be used even to deny the existence of animal minds. In the case of humans, people tended to distinguish between the rigour required by scientists in their work and our ordinary assumptions with regard to human minds. There was already a science of human psychology, dealing with fields such as cognition, emotion and will. The positivist psychologists did not normally deny the existence of these faculties. Instead they had strong views about how these faculties could be scientifically studied. In the case of animals there was at the time no discipline of psychology and thus no protector of animal minds.

Rollin discusses this scientific situation thoroughly:

> Why, then, did the study of animal consciousness vanish from mainstream psychology and biology during the early twentieth century? ... [The] explanation seems to be found in some major valuational changes – or at least emphases – which ushered in new philosophical and methodological assumptions that shaped the form of science. The major social value informing these changes, which affected a great deal of late nineteenth and early twentieth century culture, was a reductive one – the emphasis on sweeping away of frills, excesses, superfluities.
>
> (Rollin, 1990a, p. 386)

The behaviourism of Watson, Rollin points out, benefited from other values besides reductive positivism. In many of his writings Watson promised great advances for psychology. The abandonment of consciousness could put psychology on a par with physics; it would create a science that could transform humans and society, 'reform criminals, change anti-social behaviour, and create a better world' (Rollin, 1990a, p. 387).

Rollin goes on to say:

> One might be inclined to wonder how the denial of consciousness in science could possibly withstand ordinary common sense, which ... takes it equally for granted that humans and animals are conscious. The answer is simple: Ordinary common sense did not much care what science believed.
>
> (Rollin, 1990a, p. 389)

However, although science did not much alter ordinary conceptions with regard to humans it did so to some extent during the first part of the 20th century in the case of animals. The scientific denial of animal minds could exonerate certain people – the ones who were dealing with livestock for the purpose of human consumption – from responsibility with regard to creating animal suffering. If animals do not have minds there is no suffering to care about. The same idea indeed also exonerated the scientists who kept animals in order to experiment on them without much considering the state they were in.

Positivism and behaviourism maintained their dominance for several decades. One dissenting voice, Burt, commented on this in the following way:

> Nearly half a century has passed since Watson (1925) ... proclaimed his manifesto. Today, ... the vast majority of psychologists ... follow his lead. The result, as a cynical onlooker might be tempted to say, is that psychology, having first bargained away its soul, and then gone out of its mind, seems now, as it faces an untimely end, to have lost its consciousness.
>
> (Burt, 1962, p. 229)

Positivism has now lost its grip on most scientists. Prevailing theories of science (see, for instance, Kuhn, 1962; Feyerabend, 1975) are different and

have convincingly demonstrated the weaknesses of the simplistic positivist programme. It is now commonly recognized that all sciences, including physics,
require some abstract presuppositions in order to get off the ground. Not all
scientific data can be based simply on observations. When such data are based
on observations they must be interpreted according to theories and principles
and these theories and principles are not in themselves observable data. The
assumption of the existence of minds can be supported for the simple reason
that data can be organized much better with the aid of mental categories than
with the aid of purified somatic or environmental categories. It is clearly much
easier and more elucidatory, for instance, to explain human actions in terms
of beliefs, emotions and desires, than to explain them in terms of complicated
sets of inputs, without recognition of any internal mental make-up.

Another crucial fact is that the majority of humanistic disciplines (such as
history and comparative literature) have never adopted a positivistic view.
Historians and aestheticians, for instance, have always presupposed the
human mind in all its richness. Theories in the humanities, such as phenomenology and hermeneutics, presuppose a rich world of human consciousness
that has to be interpreted in terms such as goal-directedness and meaningful-
ness. These theories are now much better known in Western psychology and
sociology and have had an impact on these disciplines.

When it comes to animals important extra-scientific movements have been
crucial. The animal welfare and animal rights movements have been successful in reminding us about brute animal suffering. Hardly anyone has, when
forced to think about it, denied that at least the higher-level animal can suffer
and that we are continuously, as scientists, as livestock keepers, as zoo man-
agers, etc., causing animals suffering. The discussion following the initial
outcry from the animal welfarists has led to more research and this research
must per definition presuppose the existence of animal minds.

But things move fairly slowly in the scientific world. As has been noted by
McMillan and Rollin (2001), little has happened so far in veterinary science. In
a survey that they made the following was found. A search of the Veterinary
Information network database for English-language articles related to clinical
veterinary medicine published in the three preceding years that included the
terms mental states, mind, well-being, emotional well-being, emotional stress
and other similar terms, yielded only four articles altogether. Similarly a review
of ten current veterinary textbooks for information on the mind and mental
states indicated that of 1231 chapters only eight specifically addressed mental
states. Of these, six addressed pain.

The real change has come in that part of animal science that has been
influenced by the cognitive sciences (e.g. comparative psychology, cognitive
ethology and the neurosciences). Here a whole body of new literature has
emerged which may be called cognitive animal science. Although the focus
here is precisely on cognition, this literature also comprises texts on feelings,
preferences and dislikes as well as suffering.

In human medicine this development has been much faster. There is also
more integration in the human case between basic work on mental phenomena, in particular well-being, and clinical work than in the case of animals. The

measurement of the quality of life of humans is now a flourishing subject, both as a theoretical and a clinical one. Observing this, McMillan and Rollin in fact recommend the introduction of a mental health discipline in veterinary medicine.

McMillan and Rollin (2001), indeed, also recall the veterinarian's oath where veterinarians swear to use their 'knowledge and skills for the benefit of society through ... *the relief of animal suffering*' [my italics] (p. 1723). This oath clearly presupposes the existence of animal suffering. From this follows that society can expect veterinarians to be the advocates and protectors of animals, not only as organisms but also as minds.

Arguments for the Existence of Animal Minds

Although the existence of animal minds is now almost universally taken for granted (some exceptions – Bermond, 1997 – still exist, see below in this chapter), there is still a case for considering the arguments for this. On what can we ground our assumption that animals have minds? The question is also pertinent for the reason that we believe that different animals have minds of different kinds and different development. So how can we explain these differences and to what extent can we sustain our views concerning them?

I will take as my starting-point Roger Crisp's (1990) illuminating list of arguments for assuming the existence of animal minds. Crisp claims that the psychology of human beings and that of animals form a unity. 'Animals are not unconscious, nor are their minds entirely beyond our understanding' (p. 394).

First Crisp discusses the *Other Minds Argument*. This argument is negative: it says that there is no reason to doubt the existence of the minds of animals. Any reason that we have to doubt the existence of minds in animals we also have with regard to humans. The positivist reason says that we cannot observe the minds of any other being than ourselves. Therefore the radical conclusion is that only *my* mind is real, i.e. we end up in solipsism. However, few people end up in such extreme solipsism. We assume that our nearest and dearest have feelings and thoughts. So why should we stop at the species border? Why should we not accept that our dogs and cats can feel and think?

A common argument, cited by Crisp, to vindicate the species border is to point to the fact that only humans have language. Only humans can express linguistically the idea that they are in, for instance, pain. The dog and cat, it is normally claimed, do not have such resources.

This is, indeed, a debated issue. Many theorists hold that many animals in fact have rather complicated languages. But be that as it may, the argument from language has little force. There are many humans, including infants and the seriously mentally disabled, who do not possess language (despite Descartes's contention). Following the initial argument we should therefore also doubt that these people can have a mental life, even that they can be in pain. This sounds absurd. We normally claim that we can 'see' that the infant and the mentally disabled are in pain. There are other ways of expressing pain (as well as of expressing other mental states) than using a conventional language.

This brings us to the next argument for the existence of animal minds, viz. the *Argument from Behaviour*. Crisp invites us to consider the following example:

> A cat enters my back yard, passing by Fido's kennel. The cat makes a run for it around the corner. Fido chases her, but she is not in the yard. Fido stands barking under the lime tree. It would be very hard to describe this example other than in the following way: the cat was frightened by the sight of Fido; Fido was excited by seeing her; the cat wanted to escape Fido and believed that she could do this by climbing the tree; Fido could not find her in the yard, and so concluded that she must be up in the tree.
>
> (Crisp, 1990, pp. 396–397)

This brief story contains a variety of mental terms: 'frightened' (an emotion), 'sight' (a perception), 'excited' (an emotion), 'seeing' (a perception), 'wanted' (a volition), 'believed' (a cognition) and 'concluded' (a cognition). In fact the story covers most types of mental states ascribed to humans. (The only major category missing is the sensations, like pain, which are of course also often attributed to animals.) Crisp's contention is that the only way we can understand this sequence of events is to use our categories for mental states. The observed behaviour is quite natural if we look upon both Fido and the cat as perceiving animals, and the cat as being frightened by what she considers to be a dangerous dog. The behaviour of the two is quite analogous to the behaviour of, for instance, a little girl seeing a big boy who is frightening her and chasing her. Such an interpretation of the animal story is also strengthened by the fact that both the cat and the dog have eyes with which they look at each other.

This last observation brings us to the *Argument from Neurology*. The brains of all multicellular animals are made of the same matter, Crisp notes. 'The fundamental characteristics of neurons and synapses are roughly the same' (Crisp, 1990, p. 397). Moreover, many animals have brains which are both absolutely and relatively at least of the size of human brains. Consider, for example, small whales, dolphins and porpoises. Again we have an analogy argument. The bodily organ that is most closely associated with mental life, viz. the brain, is similar all over the animal world (with the exception of the extremely lower-level animals). All vertebrate brains can be described in three main parts: the hindbrain, midbrain and forebrain. These parts represent in the human case various elements of mental life. There is good reason, then, to believe that the animal brains, at least those of the vertebrates, can issue in the same mental functions.

Crisp mentions finally the *Argument from Evolution*. According to the theory of evolution those animals that can adapt better to their environments will survive to a greater extent than those that cannot. Crisp claims that mental life plays a role in adaptation. It is likely that a mental development has played a crucial part in animal survival and those animals with the most advanced mental life, viz. humans, have the greatest capacity for survival.

But how, then, could a mental life help? According to Crisp mental life may involve mental images.

> If a member of x has mental images of a known predator which it has been seeing preying on other members of x and of that predator as present now in the distance, as well as the ability to compare the two, it will be better equipped to survive than an animal without these capacities.
>
> (Crisp, 1990, p. 399)

To this can be added that mental life involves cognition. It is through thinking and reasoning that humans have become so well-equipped for exploiting the resources of the world for their benefit and for handling new and demanding situations.

The argument from evolution is developed in much more detail and in a more sophisticated way by Yew-Kwang Ng (1995). His argument runs from the notion of plasticity to the notions of feeling and affect. It can perhaps be summarized in the following way. Many animals, including humans, show plasticity in their behaviour. By this is meant that a given stimulus does not trigger a fixed response. The animal instead chooses a behaviour that is functional given the situation. In order to react adequately to a situation the animal must in some sense 'observe' this situation. It is difficult to imagine, says Ng, that an unconscious organism can make such flexible choices. Thus, Ng concludes, all animals having plasticity must be conscious. (A classical analysis on this theme is performed by William James, 1901, pp. 145–182.)

And if consciousness is an evolved function, which most biologists believe, it must in some way contribute to fitness. However, consciousness as such does not contribute to fitness. In order for the organism to survive in the evolutionary competition, consciousness must affect its activities by influencing its choice. The influence of the choice, says Ng, must go via a reward and punishment system. The reward in this case is the feeling of pleasure whilst the punishment is the feeling of pain or other kinds of suffering.

But why should nature need plasticity and consciousness? The answer, says Ng, is that in complex situations evolution does not know in advance what should be the right activities. For sufficiently advanced species that deal with sufficiently complex situations, it is more economical to spend some resources to provide the organism with a mind that can be aware of its environment and decide what the best thing to do is. (See Ng, 1995, p. 268.)

Crisp's arguments taken together (and improved by Ng) seem to have a lot of force. It must be difficult to *accept fully* the diverse and complex mental life of humans and *completely deny* the existence of mental life in, at least, higher animals. This is particularly so if we include such humans as lack language and other means of communicating about their 'inner' states. In both these cases and the animal ones we must rely on our knowledge about the behaviour and neurology of the living beings in question.

However, to say that animals have some mental life is not to say anything specific about what *kind* of mental life they have. In order to do so we must tell true stories about complex animal behaviour (for instance of the Fido–cat kind) or set up experiments, for example in order to measure the perceptive or intellectual capacities of animals or measure their preferences with regard to behaviour or outcomes. Some of this has been done (see, for instance, Dawkins, 1990) but most remains to be done. For my purposes in this book I shall confine myself to that part of mental life that is entailed in well-being and ill-being (see later sections).

A Rival Argument for the Existence of Animal Minds

I will now scrutinize a rather different argument for animal mental life presented by John Dupré (1990). This argument is inspired by Ludwig Wittgenstein (1953) and Gilbert Ryle (1949). At first sight it could be viewed as supplementary to Crisp's arguments and as supporting his general idea. On the other hand Dupré completely distances himself from two of Crisp's arguments, viz. those that are based on reasoning from analogy. Dupré denies the possibility of using an argument from analogy to defend the assumption of animal minds and he therefore presents a completely different argument.

Dupré denies the possibility of an argument from analogy when this argument is solely built on one case, viz. the subject's own case and the fact that he or she has a fully fledged mental life.

> An inductive argument based on the observation of one case to a generalization over a population of billions is hardly deserving the title 'argument'. The reason that we do not accept inductive arguments based on a single instance is that we cannot, in general, have any reason to suppose that the observed case is typical.
> (Dupré, 1990, p. 431)

Dupré takes as an example of unreasonable inductive reasoning the following. Suppose we have in front of us a box that emits sounds of various kinds. In fact, the object is an ordinary radio. If we used this instance as an argument for saying that all similar boxes could emit such sounds we would be in error. There are a lot of similar boxes that are not radios and cannot emit sounds.

But our failure to make such an inference, says Dupré, does not invalidate our idea that other humans than ourselves have minds or that other animals have minds. We have misunderstood the nature of minds, he claims. Or rather, we have misunderstood the logic of the mental language, as Wittgenstein and Ryle would have expressed themselves. The mental words do not refer to an 'inner' experience, at least not solely. We have for 400 years been trapped in the Cartesian view of the mind, where the mind is a special 'substance' locked inside the brain but quite distinct from the brain substance itself. As Ryle says, the mind is not a ghost-like entity in our bodily machine. The mental words do not refer to a substance or an entity at all. For Ryle they refer rather to behavioural dispositions. To be in love, says Ryle, means to be disposed to do certain things. Likewise to be in pain means being disposed to certain behaviour. Wittgenstein expresses himself in similar ways but perhaps not as clearly as Ryle. Dupré summarizes:

> Wittgenstein approaches meaning through a consideration of what it is to explain meaning … So in the present context we need to think of what is involved in explaining the meanings of mental terms; and an essential part of the answer, surely, is behaviour expressive of the mental states in question.
>
> (Dupré, 1990, p. 432)

For a more comprehensive treatment, see also Dupré (2001, 2002).

When we teach and learn the meaning of the term 'pain', says Wittgenstein, we observe ourselves and others in various painful situations. We see what happens and what we do in such circumstances. For instance, we make grimaces, we cry and we move the aching limb or point to the part of the body that is hurt. All such observations make up a catalogue of behaviour that together constitutes at least part of the meaning of the word 'pain'. The behaviours become the criteria for the correct application of the word.

Since such behaviours constitute part of the meaning of 'pain' there is much less of a problem in applying the criteria to other people and to animals. Other people cry, make grimaces and move about in apparently irrational ways and so do animals. The dog and the cat can whine and jump about, and we interpret these behaviours as expressions of pain. But it is true that the pain behaviour of animals is not completely identical with ours. We must partly learn about it afresh. For this we have other clues than our own behaviour. We can look at the causes of their behaviour. Has the animal been hurt? Are there visible injuries, etc.? In an analysis of this kind the private nature of the mind vanishes. The privateness that has all along been the difficulty is no longer a difficulty. Therefore, we need no argument from analogy, says Dupré.

The Ryle–Wittgensteinian analysis is forceful and has had a lot of followers in late analytical philosophy. It has its problems however, and one of them is particularly prominent in the welfare context. Let me examine these problems.

We can first note that at least Wittgenstein's theory sounds like analytical behaviourism (see Watson, above). In fact, Wittgenstein has been accused of being a behaviourist (Armstrong, 1968). And if Wittgenstein's theory were to turn out as simple behaviourism we would not have moved away from the stage of the annihilation of the mental life discussed above. Now, in fact, Wittgenstein is more cautious. Pain is not to be identified with pain behaviour, pain is only *partly* to be identified with the typical behaviour performed in a situation where damage has been inflicted. Dupré says:

> The distinction between Wittgenstein's position and that of analytic behaviourism … can be found in his remark that the word 'pain' does not describe pain behaviour, but replaces it. Verbal expressions of pain thus become criteria of pain on a par with groaning; and like other criteria they can, on occasion, be disingenuous.
>
> (Dupré, 2002, p. 221)

Ryle does not say that pain is a set of behaviours; pain is a (bodily) disposition to certain kinds of behaviour.

But how much do these concessions help us? Wittgenstein and Ryle may be right in pinpointing that part of the meaning of mental words is constituted by behaviour or dispositions to behaviour. This is helpful for our communication. We can identify pain in others through their pain behaviour. Thus we escape from the private trap.

But what else is there in the meaning of mental terms such as pain? Ryle does not give us more than that pain is a bodily (presumably neurological) configuration which is such that a certain behaviour issues. And Wittgenstein is rather silent about the nature of the 'inner' world. Thus, in sum, we seem to have come out of the private trap without finding an entrance in again.

This is particularly a problem for the theory of well-being. Well-being was not a prominent topic in Wittgenstein's and Ryle's analyses (in spite of the fact that pain was Wittgenstein's paradigm example). To them the main question was the nature and existence of the mental and the ways we are talking about the mental; their primary concern was not that the mental was something to *care about*. But when we discuss certain aspects of the mental life under the labels of well-being or ill-being, i.e. as something that we ought to care about, then the whole perspective becomes different. When we care about a suffering person or a person in pain we are not really interested in this person's behaviour. Our aim is not to eradicate the behaviour; instead our aim is to eradicate the awful thing that issues in the behaviour. Thus the important matter analytically is to confirm the existence of the 'inner' part of pain and other suffering. And the question is whether Ryle and Wittgenstein have helped us in this task. The answer must be that Ryle has definitely not done so; he seems to have completely closed the door to the inner room. Wittgenstein has probably not closed the door; he has, however, abstained from saying anything about the inner life.

We may note that Dupré really refers to the 'inner' sensations and experiences in his own ethical reasoning:

> [T]he concept of pain fits into broader aspects of our conceptual scheme, most especially the ethical. We think that causing pain is a very bad thing, because it is a sensation that sentient beings dislike. Since lions are clearly sentient, and show every sign of disliking the experiences which I am suggesting we refer to as 'pain', we should avoid causing them such experiences.
>
> (Dupré, 1990, p. 445)

Dupré can hardly mean here that we should avoid causing these lions a certain set of pain behaviours.

But is there, then, no way of combining the insights and suggestions of Crisp and Dupré? Can we not agree with Dupré that mental words partly signify behaviour and other bodily expressions, and that we are therefore able to use mental words also in our description of other humans and animals? But can we not also agree with Crisp that we can, with good reason, assume that other humans and animals also possess the pleasant or awful feelings lying behind the behaviour? And what is so wrong with the argument from analogy?

I wish to return to Dupré's criticism of Crisp's (and many others') reasoning from behaviour and neurology. Dupré says that one cannot argue inductively from one case and infer to a billion cases. This sounds fair enough, but the question is whether this is what is done. There seems to be a flaw in

Dupré's own analogy with the radio and the alleged inference to all kinds of similar boxes in the world. In Dupré's case we cannot ascertain that all the boxes contain the same machinery as the first box that happens to be a radio. Therefore we make quite a hasty and silly inference when we claim that all similar boxes can emit sounds. In the case of humans and animals we are acting in a different way. We are ascertaining precisely that the humans belong to the same species, that both humans and vertebrate animals are built up with the same stuff, that the inner machinery (the brains) of humans and vertebrate animals has roughly the same parts and that these are wired in the same ways. Moreover, in the case of humans we have billions of people referring to their own states of mind and referring to these states as pleasant, indifferent or awful. At the same time they behave in ways that are similar to our own behaviour in the same kinds of contexts.

These arguments offer no proof, of course. As always with induction, no proof is available. The assumption that other humans and other animals have minds similar to our own remains an assumption. But if we take the existence of minds of other humans as self-evident – and this is usually the case – then we have little reason to deny completely the existence of minds in higher animals.

The Argument that Only Anthropoid Apes and Dolphins Can Have Emotional Experiences

So far I have presented a number of reasonings to the effect that 'higher' animals can have feelings and suffer. But which are these higher animals? Are we talking about all vertebrates and possibly some further kinds or are we just talking about our closest relatives, the anthropoid apes?

An author who takes a forceful stand in this debate is Bob Bermond (1997). Under the title 'The myth of animal suffering' he argues that the claims of many animal welfare theorists are for the most part unwarranted. His conclusion is that 'emotional experiences of animals and therefore suffering, may only be expected in anthropoid apes and possibly dolphins' (p. 125). Bermond reminds us that pain and suffering are conscious experiences. Thus the question at issue ultimately concerns whether animals have a consciousness. In order to decide this issue Bermond invites us primarily to consider modern neurological research. For the sake of argument, I will follow Bermond in some detail here. This does not mean, as will become salient later on, that I find his position strong. It means instead that his points can function as an introduction to what modern neuroscience has to say about animal minds.

First, Bermond accepts that many animals can perform fairly advanced learning procedures. They can learn to overcome barriers and find solutions to problems. But even the most complicated of such procedures in humans, such as the application of the grammar of one's mother tongue or the solving of mathematical problems, can take place entirely unconsciously. Thus the existence of advanced cognitions is no argument for the existence of consciousness, according to Bermond.

Moreover, Bermond argues, humans as well as animals can perform 'emotional behaviour' without any conscious emotions being present. In particular, there can be pain behaviour without any sensory experience. This has been demonstrated in patients with spinal cord lesions. Such patients can react to an attack or an injury without experiencing anything at all. Behaviour is often first unconsciously regulated. When a human being touches something hot, such as a stove, there is an immediate reaction: the person's hand is drawn away. However, the conscious pain comes much later. The processing system for the conscious perception of sensory information, says Bermond, is relatively slow in comparison to the speed with which the information exercises control over motor responses. The conclusion to be drawn from such observations, according to Bermond, is that behaviour typically associated with feelings such as pain cannot be used as an argument for assuming the existence of conscious emotions in animals. (See Bermond, 1997, p. 131.)

What, then, about neurology? Do not all mammals have similar brain structures? Can we not, then, use a neurological argument from analogy? Here Bermond directs his most forceful attack. He argues that the prefrontal cortex is the part of the brain that is mostly responsible for the emotional life of human beings. This has been demonstrated, he says, mainly during the practice of frontal lobotomy in psychiatry. This practice has now largely disappeared in modern clinical psychiatry but a number of studies were performed during the period when it was in use.

Patients who were lobotomized showed a general emotional numbness after surgery. People who had been worried ceased to be so. Paraplegia patients showed a sudden absence of depression and guilt. Moreover, lobotomy was for a time used as a pain-killing method for non-psychiatric patients. 'The pain, which before an operation was so overwhelming that it was permanently the centre of the patient's attention, is no longer experienced as annoying after the operation' (Bermond, 1997, p. 134). Thus, although the pain could be felt somewhat also after the operation, there was no *suffering* (or reduced welfare). Bermond draws the conclusion that the presence of the prefrontal cortex is a prerequisite for the occurrence of emotional experience.

But do not all mammals have a prefrontal cortex? Bermond claims that although most mammal species have a prefrontal lobe it is only in higher apes that it is well-developed. Moreover, according to Luria (1980) some parts of the prefrontal cortex are specifically human. Thus, argues Bermond, outside the anthropoid apes (and perhaps the dolphins, which are, however, not mentioned in the specific discussion about the prefrontal lobe), the mammals represent a grey zone with respect to the question of emotional experience.

What is the strength of this series of arguments? Bermond makes certain crucial points. First, complex cognition can exist without consciousness. Second, not all sensational or emotional behaviour need presuppose consciousness. Third, there must be some neurological make-up to make consciousness claims plausible. Most theorists would agree with these points. But what follows from these general points with regard to mammals?

It is obvious that there are immediate pain reactions that do not reflect conscious pain. There are probably also other immediate reactive behaviours, such as might be interpreted as 'fright-behaviours' (avoidance or 'freezing' reactions), that occur before any conscious fear is present. But one can wonder how this observation can be applied to such emotions as, for example, grief, guilt, and boredom, where there are hardly any immediate responses – in the human case – which are not accompanied with strong subjective feelings.

The general argument from neurology is, I think, strong. As long as we believe that the mental life of humans is almost completely dependent on brain and other neural functioning – the somatic addition being certain endocrinological functions – then we should in consequence require similar structures in the animals to which we ascribe a mental life. The alternative would be to require a similarly complex structure of a completely different kind in other animals that could plausibly account for conscious experiences. The demonstration of such structures has not been performed.

But what force does the argument from lobotomy have? What it shows is that the prefrontal cortex has a crucial role for a person's sensational and emotional life. But it does not show that the prefrontal cortex is a necessary condition for consciousness. Even in the paradigm cases on which he relies, Bermond presupposes that there is consciousness after lobotomy. He says that the pain, after the lobotomy, is no longer annoying. But the pain is still there. The world is consciously interpreted as it was before. The difference from before is that this interpretation does not give rise to as rich an emotional life as before.

Therefore the argument from lobotomy does not have the general force that Bermond intends. The prefrontal cortex is in no way a prerequisite for consciousness in general. In particular, conscious perception, as also Ng argues, which is a prerequisite for plastic reactions to the outer world, needs no prefrontal cortex at all.

Bermond's remaining strong point would be the following: although some sensations and emotions will still be there after lobotomy, they will not constitute suffering. Thus, for the theory of welfare we need not bother about animals without a prefrontal cortex. What follows? Most mammals indeed have a prefrontal lobe. Bermond claims that it is only in the higher apes and the dolphins that such a lobe is highly developed. But the leap is far, if the conclusion from this is that no animals, apart from the higher apes and dolphins, can suffer from negative sensations and emotions. What evidence do we have for saying that only such pieces of the prefrontal lobe as are exclusive to humans and the higher apes are the ones that are responsible for sensational and emotional suffering? As far as I understand we have no such evidence.

Moreover, it is doubtful that Bermond is right, given empirical evidence, that pain does not constitute suffering among lobotomized patients. Roger Trigg (1970), who is in fact quoted by Bermond, gives a much more complex account. Trigg notes that some of the lobotomized patients complain less about their pain after the operation. The pain is, however, in most cases still there and it is much disliked by the subject. What seems to have happened, according to Trigg, is that the patients have partly lost their capacity for interpreting the pain and hence *worrying* about it. They are less anxious about their future but they still suffer from the pain, says Trigg.

We ought also to note that people can suffer from other feelings than pain. There is lot of clinical evidence that outbursts of anger, for instance, are common among lobotomized patients. Some patients, when they understand that their personality has been transformed by the operation, can become furious about this. Panksepp (1998, p. 79) underlines this point for the animal case: 'While frontal lobe lesions made animals more placid, they also exhibited strong tendencies for simple-minded emotional outbursts when thwarted'. Thus Bermond, in arguing for his case, forgets about the breadth of the notion of suffering. Suffering from pain is not the only species of suffering.

A more important and general point, however, is that Bermond is not representative of the opinion of modern advanced neuroscientists. It is easy to show that Bermond in his article completely neglects the wealth of findings about the neurological basis for animal emotions. I will return to a brief review of this in a later section.

A Pharmacological Argument for the Existence of Suffering in Animals

But let me first recapitulate in some detail a reasoning that runs in the directly opposite direction from the argument from lobotomy. This is an argument presented by A.N. Rowan (1988). Rowan recapitulates some results of pharmacological research where it has been found that many species besides mammals have receptors for benzodiazepine, which are the neural substrates for most anxiogenic agents today.

The starting point is human anxiety, an extremely unpleasant mood which is a central ingredient in much mental illness. Anxiety is in some circles taken to be a specifically human state of mind. Indeed, it has a central place in existential and phenomenological thinking as a mood presupposing a concept of the self and an awareness of one's mortality.

However, if one relaxes the requirements on anxiety a little, one is, Rowan says, likely to find a mood in many animals that resembles human anxiety, still remaining distinct from the emotion of fear. (The latter, as an emotion, has an obvious object. The animal fears a predator or a human being, for example.) Experts tend to describe this state in the following behavioural terms: motor tension, autonomic hyperactivity, apprehensive expectation, vigilance and scanning.

Why should this animal state, defined in behavioural terms, be labelled anxiety and be supposed to consist of conscious feelings? The reason, Rowan says, is that animals in an 'anxiety state' respond in much the same way to the benzodiazepine drugs as humans with anxiety do. Their tensions and hyperactivity are curtailed and they seem to be much more at ease. Indeed, the basic pharmacological research was, as is usual, performed on animals. When the apparently tranquilizing results on animals (without there being any serious side-effects) had been observed, the decision was taken to extend the use to humans with anxiety.

In this research a specific type of receptor for benzodiazepine was discovered in the mammalian central nervous system. It could thereafter be shown by Richards and Mohler (1984) that there is a single neurochemical substrate explaining the anxiolytic properties of alcohol, barbiturates and benzodiazepines. It has furthermore been shown that certain compounds that bind to the benzodiazepine receptor cause anxiety in humans. These compounds, when administered to humans, caused intense inner strain and excitation, increased blood pressure and pulse as well as restlessness. When they were administered to some primates they caused similar physiological and behavioural changes. (See Rowan, 1988, p. 140.) All this evidence points to an occurrence in mammals of an emotional state analogous to human anxiety. Indeed, the benzodiazepine receptors were not only found in (17) mammals; they were also found in three species of bony fish.

How is this argument from pharmacology to be assessed? The strength of the argument lies in the fact that we have both behavioural and neurological evidence. We find animal anxiety-behaviour that is similar to human anxiety-behaviour. We find the same neurophysiological event causing these behaviours in the human and the animal case. Moreover, we find the same pharmacological substance relieving the human person from his or her feeling of anxiety (as well as the associated behaviours) and eliminating the anxiety-behaviour of the animal. From here we draw the conclusion that the animal may also have a feeling of anxiety, resembling the human feeling of anxiety. It is difficult to see that we can design any more forceful argument in the discussion of animal minds.

Some Insights in Contemporary Neuroscience (Panksepp)

Jaak Panksepp (1998) has made an admirable attempt to summarize the present standpoint in neuroscience with regard to human and animal affection. Panksepp makes several crucial statements about the existence of animal conscious affection that go in a direction quite contrary to Bermond's and in favour of the assumption that most animals (at least all vertebrates) can have subjective experiences. (A terminological word of caution must be offered here. I will in these paragraphs use the terms 'affective', 'subjective' and 'emotional' states without particular discrimination. Some important distinctions that can be made in philosophical psychology between sensations, emotions, moods, etc. will be made in my constructive reasoning in Part III.)

First, Panksepp (1998) emphasizes the remarkable similarity in the organization of the brains of all mammals. A rat brain, for instance, is organized in a way quite similar to a human brain. Second, Panksepp underlines that, although certain areas of the brain are more involved in emotion than others (viz. the subcortical regions), ultimately all brain areas participate in emotions to some extent (p. 57). (This goes against Bermond's simplified idea about the frontal lobe as *the* site of suffering.) Third, he explains the existence of emotional systems in an evolutionary fashion. These systems serve adaptive functions. They help 'organise and integrate physiological, behavioural and psychological changes in the organism to yield various forms of action readi-

ness' (p. 303). The emotional centres of the brain, he says, have played a powerful role in synchronizing neural events to co-ordinate human and animal
behaviour in response to survival problems: 'to approach when seeking, to
escape from fear, to attack when in rage, to seek social support and nurturance when in panic' (p. 303). Thus, fourth, it is, Panksepp says, self-evident
to most observers that animals have subjective experiences. It is apparent in
their outward behaviour, it is apparent from their responses to certain triggers
or otherwise motivating conditions. Other compelling evidence comes from
neuropharmacology (see Rowan, above). Behavioural changes in animals can
predict human subjective responses, and brain simulation studies show that the
subjective responses of humans are completely parallel to the behavioural
responses in animals.

In support of these general and sweeping statements Panksepp provides
detailed information about the organization of the mammalian brain and what
we know so far about the centres or 'circuits' of affection and subjective experience in general. It goes far beyond the purposes of the present book to recapitulate this research. Let me only focus on some basic points.

The mammalian brain is divided into three major parts, the basal ganglia
(or the reptilian brain), the limbic system and the neocortex. The basal ganglia
organize some fundamental aspects of instinctual motor capabilities in animals,
in particular the behavioural routines that, for instance, reptiles still exhibit.

The limbic system that surrounds this reptilian core consists of several
parts in its turn, for instance the amygdala, the hippocampus and the hypothalamus. This is the part of the brain that is mainly responsible for the emotional life of animals and humans. According to Panksepp (1998), the limbic
system generates the basic emotions that 'mediate various prosocial behaviours, including maternal nurturance, associated caressive behaviours, separa-
tion distress vocalizations, playfulness, and various other forms of competition
and gregariousness' (p. 71).

The basal ganglia and the limbic system are quite similar in all vertebrates.
This holds also for many parts of the neocortical areas. The neocortex is
responsible for the transformation of sensations into perception and concept
formation. It is thus the centre of our rational abilities.

Modern neuroscience identifies some major affective neural circuits, which
are genetically coded and common to most vertebrates. Panksepp refers to
four such circuits and calls them the seeking, fear, panic and rage circuits. The
seeking circuit organizes stimulus-bound appetitive behaviour and self-stimula-
tion. The fear circuit organizes stimulus-bound flight and escape-behaviours,
the panic circuit stimulus-bound distress vocalization and social attachment,
and, finally, the rage circuit stimulus-bound biting and affective attack. In addition to these primitive emotional circuits there are more special systems that
are engaged at particular times in the lives of the animals. Panksepp identifies,
for example, the sexual lust, care and play systems. A lot of evidence has been
found for the existence of all these circuits. Among other things they can be
evoked by localized electrical stimulation of the brain. In the case of humans
relevant subjective experiences have been reported and in the case of animals
relevant bodily reactions and behaviour have been reported.

Panksepp underlines the enormous physiological and neurochemical complexity of the affective systems. All parts of the brain are involved and contribute. However, there are physical sites in the brain that seem to organize the subjective emotions of humans and animals, and most of these sites are in the limbic system, for instance in the amygdala, the hippocampus and the hypothalamus.

Knierim *et al.* (2001) rely on Panksepp's analysis in their attempt to summarize the state of the art of animal welfare science. They consider it crucial to use modern neuroscience in order to find reliable indicators for emotional states. Two things are particularly stressed by them. First, neuroscience (together with other findings) shows that animal affective life is not only negative. Also relatively primitive vertebrates have positive affective centres: the seeking, lust, care and play centres. Little attention has been paid, however, to the positive part of the subjective life of animals. Second, Knierim *et al.* (2001) insist on the impossibility of equating any specific neurological or other functions with the welfare of animals. Welfare, they say, is a multidimensional concept and it refers to evaluative dimensions captured by the ideas of functioning *well* and feeling *well*.

13 On Quality of Life in Animals

A Terminological Loan from the Human Context

After this digression on animal minds I will turn back to a feelings-based theory of animal welfare, in this case a theory that is framed in terms more frequently used in the human case, viz. 'quality of life'. As mentioned earlier in this book, 'quality of life' or 'QoL' has become a fashionable term in the human context, in particular in medicine and social affairs. 'QoL' as a term in human health-care research has almost exclusively replaced terms such as 'well-being' and 'life-satisfaction'. The latter terms are frequently used, however, for the further analysis of QoL.

In animal and veterinary science, as we have seen, the situation is largely different, although a few authors, including Fraser *et al.* (1997), sometimes use the term 'quality of life' together with 'welfare'. The favoured term is 'welfare'; sometimes 'well-being' occurs. The interpretation of 'welfare', as we have seen, varies greatly. The major interpretations are coping, natural living, and absence of suffering or positive well-being, where the latter is interpreted in feeling terms.

In an interesting article Franklin McMillan (2000) recommends the introduction of the term 'QoL' into animal and veterinary science and he proposes a specific interpretation of it. He suggests that QoL should refer exclusively to mental states of the animal. I will summarize McMillan's proposal. (My quotations are, unless otherwise stated, from McMillan, 2000.)

McMillan has made a preliminary survey of the literature and found that there is some, though sparse, use of the term QoL in veterinary science. His major conclusion, however, is that the term is sometimes used but never defined. Indirectly, though, one can find that it is often equated with health status and used to compare the efficacy of different medical treatments. This squares well with the use of the term in medical research.

McMillan sets himself the task of introducing a well-defined QoL concept for animal and veterinary science. He starts off with the bold statement: 'quality of life refers to a state of mind; it is a conscious, subjective, mental experience' (p. 1905). This statement is bold since when QoL is used in the human medical context this is not a self-evident choice of interpretation. Most instruments designed to measure QoL in medical research are multi-layered and contain both so-called objective and subjective criteria. Consider, for instance, the EuroQol instrument that covers the following domains: mobility, hygiene, main activities, pain/discomfort and anxiety/depression. (See Brooks, 1996.)

Consider now the next step in McMillan's reasoning:

> Affect (subjective feelings) plays a preeminent and, I propose, exclusive role in all interpretations of QoL in animals. For sentient animals emotions appear to be a relatively constant experience. Human and non-human animals seem to experience affect, to some extent, during all of their waking life, and all affect seems to have a hedonic quality (i.e. it is either pleasant or unpleasant).
> Therefore, affect contributes pleasantness or unpleasantness on a continual basis to personal experience.
>
> (McMillan, 2000, p. 1905)

And McMillan goes on to say: 'Affect appears to play such a central role in QoL that it can be regarded to be the single common denominator for all factors that influence QoL' (p. 1905).

There are at least two things to disentangle here. McMillan wavers between proposing a definition and making observations about causal influence. On the one hand he *proposes* that affect should play a central role in the interpretation of QoL. Here he makes a stipulative definition. On the other hand he later *observes* that affect appears to play a central role in influencing QoL. Here he is making an empirical observation. I take his major intention, though, to be the forming of a proposal of a definition of the technical concept of QoL.

The other question to be asked concerns McMillan's concept of affect. He first identifies affect with subjective feelings, whereby he seems to include all kinds of feeling (including sensations, emotions and moods). On the other hand he explicitly talks about emotions as if they constituted the whole territory of affect. My general interpretation here (supported by McMillan's further development of the concept) is that affects by McMillan are not understood exclusively as emotions in the narrow sense (as feelings directed towards objects) but as also including sensations and moods. McMillan presumably wants to incorporate pain (sensation) and objectless anguish (mood).

McMillan's next step is to emphasize the individual nature of QoL.

> In people, QoL is determined by the nature of the individual's experiences and by the values and meaning that the person attaches to those experiences. Individual preferences, values and needs, which derive from the individual's unique genetic make-up and learned experiences, lead each individual to assign

> different values and priorities to different aspects of life. For example, one
> individual may value social companionship over physical health, whereas another
> individual may have the opposite values.
>
> (McMillan, 2000, p. 1905)

McMillan goes on to say that there is now a general consensus that QoL should be assessed from the perspective of the individual, incorporating that individual's values and preferences. This contention is not true. In standard measurements of human QoL certain domains have been preselected and the subject cannot change these domains. In the case of some, but not all instruments, the subject can make a general overall estimate of his or her satisfaction with life. The completely subjective interpretation that McMillan endorses does not represent a common interpretation.

McMillan makes a further important statement about animals in this regard. He clearly extends the idea that QoL must refer to the individual's evaluation to cover also the animal case. He admits the salient methodological problems that this view entails. 'Nevertheless, the principle of assessing QoL from the viewpoint of the patient remains the goal. Fortunately, innovative research methods are providing new insights into animals' personal preferences and values' (p. 1905).

McMillan's Concept of Quality of Life

McMillan's more detailed concept of QoL has two major components. First, he says that QoL covers a dimension of affective states called comfort and discomfort. These states, as he says, constitute a continuum of feeling ranging from comfort to extreme discomfort. 'The term "comfort" relates to the experiential mental state of ease and peaceful contentment with no (or minimal) discomforts' (p. 1905). Discomfort relates to any unpleasant or disagreeable feeling, or any negative affect.

McMillan connects his notion of discomfort to the notion of need in the following way: 'We can say that for an animal, a need is something that would result in discomfort or a threat to an animal's life or well-being if it were unfulfilled. Needs satisfaction, therefore, lessens discomfort or threats to well-being, thereby assuming a prominent role in QoL' (p. 1906).

The second component in McMillan's concept of QoL is a dimension of pleasure. He says that pleasures of physical origin contain such things as physical contact and gustatory pleasures. Pleasures of emotional origin include social companionship and mental stimulation.

Thus he summarizes:

> Quality of life is a multidimensional, experiential continuum. It comprises an
> array of affective states, broadly classifiable as comfort–discomfort and pleasure
> states. In general, the greater the pleasant and lesser the unpleasant states, the
> higher the QoL. Quality of life is a uniquely individual experience and should be
> measured from the perspective of the individual.
>
> (McMillan, 2000, p. 1907)

A crucial element in McMillan's thinking, expressed in a later article (2003), is that elements along the QoL-dimension may but need not have anything to do with survival value. Here, McMillan distances himself from, for instance, Broom for whom survival value or coping is the core element in welfare.

Analysis of McMillan's Concept of Quality of Life

McMillan clearly aligns himself with those welfare theorists who, like Dawkins and Duncan, place welfare on the feeling scales of animals. To have welfare for them is to have pleasant feelings, and in particular, not to suffer. Similarly, to have a high degree of QoL for McMillan is to have a high position on the comfort–discomfort scale or on the pleasure scale.

What are the differences in McMillan's approach? One special feature of course is the terminological one. McMillan chooses the term now fashionable in human health care, viz. QoL. This choice may have the advantage of not presupposing any of the connotations earlier associated with animal welfare or animal well-being. It is interesting, though, that McMillan in contradistinction to most human QoL theorists refers solely to mental states.

McMillan also emphasizes his universal approach. He notes that the animal welfare literature focuses almost entirely on physical pain and at most pays lip service to the existence of emotional suffering. He thinks that this negligence is detrimental in the end also for the treatment of animals. He says that, contrary to the prevailing view, 'there is evidence that emotional pain may induce greater suffering than physical pain. Studies have shown that emotional factors weigh more strongly in animals' behavioral choices than physical pain' (2003, p. 185).

Another distinctive feature is that McMillan extends his notion to levels of positive pleasure. Whereas most earlier animal welfare theorists (notable exceptions being Broom and Fraser) have focused almost completely on suffering and the absence of suffering, McMillan acknowledges also the genuinely positive states. On the human side one can see both kinds of theories. Most instruments measuring QoL focus on the negative states and the absence of such states. This is natural since QoL in the human case is often largely health- and disease-related. The starting-point for the use of the instrument is normally the existence of a disease in a subject. In most general analyses of QoL, on the other hand (cf. Nordenfelt, 1994), a positive dimension is also acknowledged.

A question can be asked with regard to the two dimensions. Why does McMillan end up with two scales? Could they not be unified in the following way?

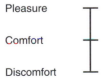

A comparison can here be made with many health scales, where the dimension runs from optimal health via minimal health to maximal ill-health. When I suggest this I presuppose that comfort is interpreted by McMillan as the absence of discomfort, i.e. that comfort exists when the subject has got rid of all pain and other suffering. Under this interpretation one can see the positive pleasure states as states over and above comfort. This means that the pleasure states are more pleasant than simple comfort.

But the two-domain proposal may warrant a different interpretation, i.e. one where comfort is not merely the absence of suffering or discomfort, but a genuinely positive state in itself. McMillan puts it this way: 'The term comfort relates to the experiential mental state of ease and peaceful contentment with no (or minimal) discomforts' (p. 1905). Perhaps ease and peaceful contentment should be understood as something over and above the absence of discomfort. But if this is so, we need an explanation of the difference between this ease and contentment and other positive pleasures. Perhaps McMillan has primarily physical pleasures in mind, such as the ones connected to eating, resting and physical (including sexual) contact.

Fraser and Duncan (1998) provide an explicit argument for the existence of different scales. They claim that negative affective states evolved in response to need-situations, where 'the fitness benefit of an action has increased, often because the action is needed to cope with a threat to survival or reproductive success' (p. 383). Positive affective states, on the other hand, have evolved in opportunity-situations where an action, such as playing, 'has become advantageous because the fitness cost of performing it has declined' (p. 383). It ought to be observed that this argument mainly concerns the genetic development of negative and positive affect, whereas my own reasoning (below and in Part III) deals with the phenomenological analysis of affect as well as with the conceptual analysis of concepts of affect.

My own proposal on this point is to regard the emotion happiness (which is close to McMillan's contentment) to be the primary dimension that runs from extreme happiness to the deepest unhappiness, via the state of minimal happiness. The pleasant sensations, moods and other emotions are things that the subject is happy about and can therefore be contained in his or her happiness. In such a proposal there is thus no need for more than one basic scale. (For a discussion about this, see Chapter 22.)

14 The Idea of Welfare as Fulfilment of Preferences

A popular idea for the characterization of both human quality of life and animal welfare is the idea of fulfilment of preferences. It is central in many accounts of human happiness (McGill, 1967; Tatarkiewicz, 1976; Veenhoven, 1984; Nordenfelt, 1994) and has been introduced into the discussion of animal welfare by for instance Dawkins (1990) and Sandøe (1996). Before scrutinizing these attempts we should first bear in mind at least three different ways in which fulfilment of preferences (or the satisfaction of wants or desires) can play a role:

1. Welfare is defined in terms of fulfilment of preferences: to say that A has a high degree of welfare *means* that most of A's preferences have been fulfilled.
2. Welfare is assessed via the person's or animal's choices in certain situations. One could say that the person or animal through these choices determines the conditions for its high degree of welfare.

The following interpretation is linked to 2 but has a different sense:

3. The person (this is hardly possible in the animal case) is given the opportunity to make an evaluative choice with respect to his or her *definition* of welfare.

The first interpretation is the most clear cut and the one that I will primarily focus on later in this discussion. Let me therefore just briefly interpret and comment on the other two. It is a popular idea in animal science, most clearly set out and practised by Dawkins, that one can determine the level of animal welfare by studying various situations where the animals make choices. This idea also enters into Broom's treatment. He says: 'The most useful source of information concerning resources necessary for good welfare is the carefully controlled preference test. As Duncan and Petherick and Fraser describe, several sophisticated methods are now available to determine the strength of preferences of animals' (Broom, 1993a, p. 23). Broom, however, also observes that preference satisfaction is not a completely reliable method: 'When an animal shows preference for a certain food or level of eating, for example, it may be mistaken in its choice and its welfare may become poor.' And he makes the analogy with humans: 'Anorexic girls have a very strong preference not to eat, to the detriment of their welfare' (1991, p. 4174). (For similar observations in the human case and possible solutions, see Nordenfelt, 2000.)

Discovering the preferences of animals is here a *method* by which we can determine the welfare of animals. Neither Dawkins nor Broom explicitly defines welfare in terms of preferences. In Dawkins's definition (1990) (see above) she says instead that 'animal welfare involves the subjective feelings of animals'.

We will be able to scrutinize also the methodological viability of preference studies in the course of analysing the viability of a preference-based definition.

The third idea with regard to preferences is different. It has to do with the basic evaluation involved in the meaning of welfare. I will illustrate for the human case. Assume that the deep-down sense of the expression 'A's welfare' is: the state of affairs that is good for A. Assume, moreover, that there are no absolute values. There is no *eudaimonia* to be discovered. Nor do we believe that the state of goodness of a man A is identical with what a consensus conference can agree upon with regard to people in general or to A in particular. Instead, what is good for A *means* what A himself considers good for A. (This can be specified and qualified in ways mentioned in Chapter 5.) This is in a way a situation of choice. A can choose a definition of his own welfare. For instance, A can insist that what is good for him is to live like a monk in the desert. Such a life, then, constitutes A's welfare.

One could then ask: how is this situation of choice to be distinguished from the one above? In the latter situation A contributes to the very definition of welfare. In the former situation A is not involved in a defining process. The choices that A makes there are only indications of A's welfare given a pre-existing definition of welfare.

But will there be a lot of difference in practice? Will a person not always choose what he or she explicitly considers to be good for him or her? Can one truly choose a concept of welfare without opting for things that are subsumed under this welfare? This is a deep question and I will not attempt to resolve it since it has only minor significance in the world of animals. With humans one can distinguish between a conscious choice of a concept and choices of courses of action. This appears difficult with animals. On the other hand, it could be argued that the conscious choice is not entirely reliable for deciding what a person considers best for him or her. It is what the person really does that is the criterion of what he or she considers best.

Let me now turn to definitions of quality of life for humans and welfare for animals in terms of fulfilment of preferences or satisfaction of wants and desires. I pick up my own definition of positive quality of life (which I equate with happiness): P is completely happy if, and only if, P wants at t that x1,...,xn shall be the case at t, x1,...,xn constitutes the totality of P's wants at t, and P finds that x1,...,xn is the case. (Observe that I distinguish between *being* happy and *feeling* happy. For a further discussion, see below and Chapter 22 of this treatise.)

Thus P has the best quality of life if, and only if, P observes that all his or her present wants are being satisfied. A similar definition applied both to the human and to the animal case has been suggested by Sandøe (1996, p. 12): 'A subject's welfare at a given point in time, t1, is relative to the degree of agreement between what he/it at t1 prefers ... and how he/it at t1 sees his/its situation – the better agreement the better welfare.'

There are many points to note about these definitions. I will first consider how to interpret them. It is crucial to see that although these are definitions of subjective welfare, they are not equivalent to the feeling-state definitions above. Welfare, in this sense (I use the term 'welfare' now to cover also the human case), is *not* identical with pleasure or any other collection of positive sensations. Welfare is instead defined as a relation between the subject's preferences and the subject's perception of the present state of affairs. There can be a close relation, however, between preference-satisfaction and positive sensations. The object of preference can be pleasure. The subject can desire to attain pleasure of a certain kind. But the desire can also be directed to concrete states of affairs, such as getting food, going outdoors or exercising.

Another relation between desire-satisfaction and positive feelings is that the satisfaction of a desire, whatever the goal of this desire, normally leads to a *feeling* of satisfaction. There is a feeling associated with the satisfaction itself (even though the goal of the desire need not involve this feeling). It is possible to require per definition that such a feeling constantly occurs for welfare to exist. Neither my own nor Sandøe's definition makes these requirements. My own and Sandøe's definitions require, however, the existence of mental states in general. Crucial concepts in our definitions are wants, desires (and other preference concepts) and perception.

Let us observe that although there is typically a close link between desire-satisfaction and positive feelings, we could find cases where a person wants something painful to be the case, and where a person tries to avoid something pleasant. An example of the former is the pregnant woman who is looking forward to the pain involved in childbirth, since it is a necessary step for her towards having a child. An example of the second type of case is the person who has taken a drug and in a clear moment detests the pleasure that the drug temporarily gives him.

Consider, second, the scope of this notion of welfare. One objection to the application of the notion to the human case is that it does not concern the past and the future. But a person (in contradistinction to most animals) may be happy also about past events and have positive hopes for the future. These ought to be included in his or her welfare, but they are not visible in the definition.

My reply is that these aspects are in fact covered by the definition. If one remembers with pleasure some past facts, then there is a match between one's present desires with regard to the past and one's present observation of what indeed happened. If one wants something to become the case in the future, then one may be happy about the present situation in the light of this future hope, for instance if one observes that the state of the world now is approaching the desired future state in the right way. This can be formally constructed in the following way. Person A wants at t1 X to be the case at t10. A judges that, in order for X to become the case at t10, then Y must be the case at t1. A judges that Y is the case at t1. Thus, A is satisfied with his or her present situation in the light of the hopes and plans for the future.

But what about sudden positive surprises (a situation that is common also in the animal case)? Assume a man who has a limited set of wants. He may be a person who lacks imagination and does not aspire to much in life. Assume now also that all his wants are satisfied. Thus he is, according to the theory, completely satisfied. But suppose that he suddenly has a feeling of great pleasure. It may be a piece of delicious food that he comes across and did not know about before. Since he did not know about it he could not desire it. But would we not say that this sudden sensation of pleasure added to his welfare?

I have elsewhere (1993) tried to handle this case within the theory. What normally happens when one is suddenly 'struck' by a pleasure is that one immediately wants to keep it. Thus, at the same time as the positive sensation appears there is a new want on the part of the subject. This want then remains satisfied as long as the sensation lasts. The appearance of the want is not trivial for the ascription of welfare. As I noted above, we may encounter instances of unwanted pleasure (for instance the pleasure of intoxication which is coupled with knowledge about the negative consequences). Therefore the pleasure in itself is not sufficient for welfare to occur. The pleasure has to be wanted for welfare to occur.

An interesting question that is brought to the fore by the case of an increasing number of wants is the following. We say that a person can be completely satisfied when all ten of his wants are satisfied at a given moment. Assume now that at a later moment this person has 20 wants and they are all satisfied. Is this person then not a happier person at the later moment? My answer is that this need not be the case. (See Nordenfelt, 1994, 2000 for discussions of this problem.)

15 Theories of Welfare in Terms of Needs

Many animal scientists talk about welfare in terms of the satisfaction of the animal's needs. I wish to scrutinize this idea. What about saying that the concept of *need* could play the role that the biological *telos* (see my discussion in Chapter 10) seems unfit to do in finding the *eudaimonia* of the animal? I will argue that the concept of need is equally deficient for the purpose, since it is closely tied to the idea of a goal, similar to *telos*. But let me defend this position by making a digression on the basic concept of need.

Towards an Analysis of the Concept of Need

The ontology of needs is mysterious and elusive. What kind of thing is a need? Is it a property of a human being? Can it be located? Is it a bodily state? Or is it something outside the human being to which he or she has some kind of relation?

Questions such as these are prompted by the fact that we have at least two kinds of locutions: A has a need for y; and y is a need for A. Assume now a situation where we say that A has a need for food. This locution is typically used when A lacks food, when there is no food in A's stomach. But if the need is something present within A, then the need cannot be identical with the food. But in the same kind of situation it is quite proper to say that food is a need for A. Here the need is explicitly equated with the food.

We have here encountered an obvious ambiguity and I think it can be resolved when we realize the relational nature of needs. I argue that 'need' is basically a relational term, more precisely a four-place predicate. This is most easily seen when considering locutions where 'need' is a verb: A needs y in order to attain G. John needs a hammer in order to repair his house. What does this locution mean? Essentially it means that John's using a hammer is a necessary condition for his repairing his house.

In such locutions we can detect the most general and clearest sense of 'need'. 'Need' here simply stands for any necessary condition for the attaining of a goal. The term 'goal' is so far not specified; it can be a goal set by a person or an animal but it can also be a goal in an abstract sense.

Needs or necessary conditions are dependent on background situations. It may be necessary for John to have a hammer in a given situation in order to repair his house. But in a different situation, where Steve does the hammering, John does not need a hammer. Therefore, needs in this general sense may vary over time depending on situational change.

Return now to our initial question about the ontology of needs and the different ways of expressing the existence of needs. The two locutions 'A has a need of y' and 'y is a need for A' should have the same analysis. They are both elliptic formulations of the following proposition. There is a goal G and a situation S such that y (or using y) is a necessary condition for A in S in order for A to attain G. More simply put: y is a need for A in S to reach G. This is the four-place expansion of the locution. We can say that A is the *subject* of the need, y is the *object* of the need, S is the *situation* of the need and G is the *goal* of the need.

But what about ontology? Where is the need? If need, as we have suggested, means 'necessary condition', then that which constitutes the necessary condition is a need. In the case of John's repairing the house, the hammer or, better, the use of the hammer is the need. In the general sense, then, a need is not a bodily state of a person; it is rather any kind of state or event, in which the person may or may not be involved, which is a necessary condition for the person's attaining a goal.

On Biological or Vital Needs

However important the basic analysis of need is, it cannot do justice to, or create an understanding of, the biological or psychological discussion about needs. This discussion is characterized by at least the following features:

1. Needs are restricted to living beings.
2. Needs are related only to goals that have some objective biological basis.
3. Needs are enumerable, and are often presented in the form of a short list of basic needs.

The basic idea here is that needs can be discovered by an inspection of the biology and psychology of a human being, animal or plant. Moreover, the assumption is that we should look for some universal characteristics which are the typical biologically or psychologically based needs.

In order to illustrate I will concentrate on one significant theory of human needs. I choose that of Abraham Maslow (1968), partly because it is a rich and substantial theory and partly because it has a central place in modern psychological theory. Maslow puts forward the following theses:

1. There is a limited set of basic human needs universal to all human beings. Maslow suggests the following five: physiological needs, the need of safety, the need of belongingness or love, the need of esteem and the need of self-actualization.

2. The basic needs are ordered in a hierarchy where the physiological needs are the most basic and the need for self-actualization the least basic. This hierarchy entails that a more basic need must be fulfilled before a less basic need can be recognized by the agent in question.

3. The basic needs are manifested in human beings as physiological and psychological drives, i.e. as tendencies within the human being to fulfil the needs.

The latter thesis constitutes an important addition to the general notion of a need, meaning merely a necessary condition for realizing a goal. Here a need is connected or even identified with a particular biological and psychological reality. Maslow assumes that for all basic needs there are physiological and psychological mechanisms that trigger off some behaviour on the part of the individual to realize a certain goal. For instance, hunger is associated with (or even identified with) a certain physiological mechanism that triggers off food-seeking behaviour. Given this idea, one can see how it has been plausible to identify needs with bodily states. The bodily states are the biological drives.

How can this influential theory of biological and psychological needs be harmonized with my previous analysis of needs in terms of necessary conditions? The answer is that it can be harmonized if we stick to our analysis of needs as relations, i.e. as four-place predicates. Consider again the locution: A has in S a need for y in order to attain G. When A on a particular occasion is hungry, then A has a need for food in order to become satisfied. If we stick to our suggestion above that the necessary condition for realizing the goal is to be identified as the need, then the food or the intake of the food is the need in this situation. This is also what I have called the object of the need.

This solution is indeed in harmony with common parlance and with Maslow's theory. When we identify needs, as also Maslow does, we do so in terms of the objects of the need-relation. We talk about our need of nutrients and sleep, as well as our need of safety, need of belongingness, need of esteem and self-actualization. All of these objects constitute necessary conditions for a person's ultimate health or satisfaction.

So where does the drive come in? It so happens that in certain, but not all, cases where a person, or indeed in general an animal, has a need then the individual's biology or psychology provides information about the need. This information has the form of an urge or a sense of dissatisfaction. Such a state is what certain biologists and psychologists call a drive.

Such a drive is clearly associated with the need-relation. I do not dispute that the drive is in some discourses identified with the need. For the sake of clarity I do not recommend, however, that we talk about the drive as the need. The first reason is that we also commonly talk about the object in the need-relation as the need. This mode of speech is more logical given the basic sense of need as a necessary condition for attaining something. Another reason is that a drive only occurs under some circumstances, viz. the circumstances when the subject *lacks* what he or she needs. One is hungry only when one *lacks* nutrients; one is thirsty only when one *lacks* water; one is afraid only when one *lacks* safety, etc. But do we not intend to say that a person or animal needs food, water and safety *all the time*, viz. also when the goal of the need is temporarily realized?

The Place of Needs in the Theory of Human Quality of Life and Animal Welfare

We can now see that the analysis of needs presented above moves us close to the idea of a *telos* for human beings or animals. The *telos* can be looked upon as the goal of a need of a person or an animal. The *telos* should be reached in order for welfare to exist; and the need should be realized in order for the person or animal to reach a desired state of affairs. Thus, in order for the need-language to be meaningful in a welfare discussion, we must ask about the relevant goal. We must ask, for instance, what close interactions with other human beings are needed for. But we must be careful in formulating the goal. If we use the need-terminology in order to clarify the notions of quality of life or welfare, then we cannot say that the goal of the need simply is to reach quality of life or welfare. In such a case we have just been moving in full circle. Our initial idea was to analyse the concept of quality of life in terms of needs. We cannot then just refer back to the concept of quality of life (Liss, 1993; Nordenfelt, 1994).

The Danish psychiatrist Anton Aggernaes (1994) is one of the theorists who have used a philosophy of need in their study of human quality of life. He has set himself the task of identifying a limited set of universal human needs. As a result he proposes a list that is simpler and slightly different from the classical Maslowian list of needs (Maslow, 1968). Aggernaes proposes the following fundamental human needs:

1. The elementary biological needs.
2. A need for warm interaction with other human beings.
3. A need to engage in meaningful activities.
4. A need for a varied and to some extent exciting and interesting life.

Aggernaes chooses the term 'objective quality of life' (QoL) when determining the degree to which an individual has realized his or her fundamental needs. He fails, however, to inform us about the goals of the needs specified in his list. He says that objective quality of life exists when these needs are satisfied. But when are these needs specified? What constitutes the goals of the needs?

For the theory to work we must then specify quality of life in other terms. This is what Torbjörn Moum (1994) does in his discussion of needs. He in fact bases the notion of need on the notion of happiness. 'If it can be shown that a person P obtains more happiness or utility from a given state of affairs A than from another possible state of affairs B, then P may be said to have a need for A (relative to B)' (p. 82). Here the goal of need is happiness. Quality of life is defined in terms of happiness, not Aristotelian *eudaimonia* but happiness as an emotion of the individual. Happiness in this feeling sense, then, is a further candidate for being a *telos* as an objective value.

Compare now with animal science need theories (B. Hughes and S.E. Curtis; see Curtis, 1987). Curtis introduces the notion of need into the animal welfare discussion. He also applies a simplified version of Maslow's need-structure. He assumes a hierarchic organization of animal need along the following lines (from lowest to highest): (i) physiological needs; (ii) safety needs; and (iii) behavioural needs.

A great deal is known about the physiological needs of agricultural animals, says Curtis. A deficiency or an excess with regard to one of the objects of these needs can constitute a stress, and an animal's responses to stressors can affect its productivity both directly and indirectly, for instance by directing nutrients from productive processes to maintenance. 'Thus for business reasons, not to mention humane ones, these most basic needs generally are not being neglected in animal production' (1987, p. 370).

Safety needs stand second in the suggested hierarchy of needs for domestic animals. In agriculture these needs are attended to less often than physiological needs, although laxity in this respect may result in injury or death. 'The fact that they are generally tolerated presumably means many producers deem them to be acceptable, but they probably are not' (Curtis, 1987, p. 371). Among threats to safety Curtis mentions weather accidents, predation and equipment hazards.

With respect to behavioural needs Curtis mentions three categories: abuse, neglect, deprivation. Abuse refers to active cruelty. Neglect refers to passive cruelty of the kind that occurs when an animal is confined. Deprivation is the sort of passive cruelty that involves the denial of certain elements of the environment that are considered less vital than the physiological and safety needs.

Barry Hughes completes this story by proposing a model of motivation for the assessment of behavioural needs. Behaviours may be assigned to categories in accordance with one of the following: (i) environmental stimuli are important; (ii) internal and external factors both contribute significantly to these behaviours; and (iii) internal factors are the chief releasers. 'Examples in the laying hen for these categories would be: (1) escape and agonistic behaviours, (2) sexual crouch and dustbathing, and (3) pecking and nesting' (Hughes, 1980. See also Curtis, 1987, p. 372.)

Hughes suggested that to ensure animal well-being, it is essential that the environment permit the animal to perform all of the behavioural patterns in category 3, and in addition it is desirable to provide for some of those in category 2. Hughes has also provided criteria for detecting what behaviour constitutes an essential behavioural need. In a case where the expected behaviour cannot be performed for environmental reasons, Hughes requires clear evidence of either frustration (such as stereotyped pacing by a hen before she lays an egg) or distortion of the behavioural pattern, for the pattern to be considered an essential behavioural need (category 3). To determine animal needs in order to provide for their fulfilment the animal must be subjected to thorough behavioural analysis.

Other authors who have proposed a hierarchy of needs for animals are Hurnik and Lehman (1985), who suggested the following levels in the hierarchy: life-sustaining needs, health-sustaining needs and attitude-sustaining needs. A further author in the animal science camp who attempts to explicate a concept of need is M.S. Dawkins (1983). Her aim is to show that the concept of 'ethological need' could be made precise and useful by relating it to a methodology now used in other areas of animal behaviour. She starts by introducing a distinction between two kinds of need: ultimate and proximate needs.

Animals need food and water in the sense that without them they die. Dawkins calls this kind of need ultimate needs: death and reproductive failure result if they are not met. Hens in battery cages do not die through being unable to dust bathe. It is still reasonable to say that a hen has a need to do so. Need is here, according to Dawkins, used in the sense of 'high causal factors'. 'The hen might have a proximate (here and now) need to behave in a certain way even though, in the unnatural environment of the cage, death will not be the result of failing to do so' (1983, p. 1197). Dawkins explains:

> In a natural environment proximate and ultimate needs will generally go hand in hand. For example, animals that experience a proximate need for food and start searching for it when their reserves are low, will be fulfilling the ultimate need of avoiding death by starvation. But in an unnatural environment such as a farm or a zoo, proximate and ultimate needs may become decoupled. Thus a bird of a migratory species kept in a zoo may have no ultimate need to migrate (human keepers will feed and care for it during the winter) but it may nevertheless have a strong proximate need to do so and this has evolved because, in nature, those that migrate survive much better than those which do not.
>
> (Dawkins, 1983, p. 1197)

This distinction is easy to understand, but strictly speaking it is not developed according to the same ground for division. In characterizing proximate needs Dawkins does not identify the goal of the need or what happens if the need is not fulfilled, like death in the case of ultimate needs. Proximate needs are such as have 'high levels of causal factors', i.e. are preceded by some strong urge on the part of the animal. This means that ultimate and proximate needs do not exclude each other. Later Dawkins introduces suffering into the context and calls suffering a 'proximate' experience. It appears then that (many) proximate needs are such that if they are not fulfilled they create suffering.

Dawkins suggests that in order to determine which proximate needs (including those which are also ultimate) are most important to animals (in the sense that not fulfilling them will create the highest degree of suffering) one could essentially follow the model of economists when they rank preferences. Other researchers have measured the strength of the preference of pigs for earth floors by forcing them to make a larger number of responses in order to be allowed access to the earth. Dawkins recommends that animal scientists use these well worked-out ways of obtaining indices of preference. 'Here is the key to discovering which of the proximate needs an animal might possess are sufficiently strong to be likely to give rise to suffering if unsatisfied' (1983, p. 1199).

In commenting on this method Dawkins makes the observation that necessities are defined by what people or animals regard as important to them at the time of making the choice, not by what is good in the long run. In other words, suffering is more closely related to proximate than to ultimate needs. (Again one can wonder about the grounds for division. Here 'proximate' and 'ultimate' seem to have to do with time and therefore exclude each other. It could very well be that suffering (and not death) is the ultimate and long-term consequence of a certain choice.)

Let me conclude the discussion about needs in the welfare context. One plausible idea for the characterization of animal welfare – like quality of life for humans – is to say that welfare (quality of life) is tantamount to the fulfilment of needs. If a need of an animal or human is compromised and not fulfilled, then this animal or human has a lower level of welfare or quality of life than if it were fulfilled. One problem with these attempts is that they have often been pursued without a close analysis of the logic of the concept of need. The most obvious neglect is that one has not seen that a need must, for reasons of logic, have a goal: *to have a need is to have a need for something in order to reach or maintain a certain goal*. The classical Maslowian theory, to which several authors, both in the human and the animal context, refer, is also surprisingly silent on this point.

Mostly, one can, however, deduce from the context what are the goals of Maslow's needs (see Nordenfelt, 1987/1995). Similarly, one can understand what Curtis and Dawkins regard as the goals of the needs by looking into what they consider to be the consequences of tampering with these needs. Not fulfilling physiological needs leads to death and not fulfilling other needs leads to suffering to some extent. These goals are crucial to make the theory of welfare or quality of life at all meaningful. The factor constituting the goal is the factor that in essence determines what the author considers to be included in welfare or quality of life.

So, what can one find in the writings of Curtis, Hughes and Dawkins with regard to the goal of needs? Curtis and Hughes refer to death, frustration and distorted behaviour if the needs are not met. Thus, welfare to them must, self-evidently, involve life itself, but also, presumably, involve well-being in some feeling-sense, and 'normal' behaviour. In the case of Dawkins suffering is particularly mentioned (apart from death), which means that she claims that well-being in the feeling-sense is the essence of welfare.

A conclusion from this reasoning is that a need-analysis of the notions of welfare and quality of life must in the end become an analysis in terms of the goals of needs. To say that welfare is the fulfilment of needs is to say that welfare is tantamount to attaining the goals of the needs, which are, as we have seen, normally described in terms of positive well-being or avoidance of suffering.

16 Theories of Welfare in Terms of Natural Behaviour

The Notion of Organic Farming

A different idea for characterizing welfare is to focus on the behaviour of the animal. I have in passing noted Broom's use of the idea of 'normal' behaviour as an indicator of welfare. I there criticized the viability of this notion since normality, in a statistical sense, cannot in general indicate welfare.

A different approach, worthy of attention, is the one where the term 'natural' replaces 'normal'. This is the approach that says that we should raise animals in 'natural' environments and allow them to behave in 'natural' ways. According to Rollin (1996, p. 10) consideration for the animal's nature implies that it should be able to express its natural forms of behaviour, such as play, and its natural forms of movement and social interaction. Rollin's moral conclusion from this idea is the following:

> I am now explicitly suggesting that the essence of our substantive moral
> obligations to animals is that any animal has a right to the kind of life that its
> nature dictates. In short I am arguing that an animal has the right to have the
> unique interests that characterize it morally considered in our treatment of it.
> (Rollin, 1992, p. 90)

A version of this idea is emphasized in the world-wide movement of organic farming. The Danish researchers H.F. Alrøe *et al.* have in a recent article (2001) presented the philosophy of organic farming and discussed its implications for animal welfare issues. As a summary of this philosophy they quote the definition given by the Nordic organic associations printed in a document issued by the Danish Ministry of Food, Agriculture and Fisheries (1999):

> Organic farming is conceived as a self-sufficient and sustainable agro-ecosystem
> in equilibrium. The system is based as far as possible on local, renewable

resources. Organic farming is based on a holistic vision that encompasses the environmental, economic and social aspects of agricultural production, both from a local and a global perspective. Thus, organic farming perceives nature as an entity which has a value in its own right; human beings have a moral responsibility to steer the course of agriculture such that the cultivated landscape makes a positive contribution to the countryside.

(Alrøe *et al.*, 2001, p. 281)

Given this ideology, entailing a systemic view of nature and a holistic view of an animal, Alrøe *et al.* find that too narrow a definition of animal welfare will not do. They are aware of and discuss the following basic conceptions of animal welfare:

1. The animal should feel well, corresponding to the concepts of experience, feeling, interest and preference.
2. The animal should function well, corresponding to the concepts of need and clinical health.
3. The animal should lead a natural life through the development and exercise of its natural adaptations, corresponding to the concept of the innate nature of the animal.

With regard to the latter concept Alrøe *et al.* claim that the animal has a genetic or innate nature that has emerged through evolution, domestication, breeding and biotechnology. But there is also naturalness or integrity which expresses 'the organismic harmony that can be broken by significant and fast modifications of the natural ancestral form by way of operation, medication, breeding and biotechnology, including genetic engineering' (2001, p. 284).

Given a broad definition of animal welfare (a definition that is not limited to animal function but includes feelings or natural life, or both) it is possible, Alrøe *et al.* claim, to hold the ethical view that livestock welfare should not be limited to ensuring that basic needs are satisfied.

Rather, it should ensure that an animal lives a richer life with the opportunity to express a greater part of its natural behavior (e.g. play and social behavior). It is in this context that organic farming's emphasis on the naturalness of the production system becomes crucial, for example as it is expressed in the requirement for access to open-air and grazing areas.

(Alrøe *et al.*, 2001, p. 290)

There are two things to be separated in Alrøe *et al.*'s claims: first, their emphasis on natural environments and natural behaviour; second, their insistence on broad definitions of welfare. I will return to the latter idea in the following section.

Consider first the concept of natural life. The analysis proposed by Alrøe *et al.* and others (including Rollin) is that every animal has received through evolution an innate genetically coded nature. According to this nature the animal behaves in certain ways, unless it is prevented from doing so. It seeks food in certain ways, it provides for its shelter and breeding in certain ways, etc. If the animal is allowed to behave in this predisposed way then the animal will be happy. The natural behaviour presupposes a 'natural' environment. In order to behave in a natural way the animal must have a background in terms of an environment that has not been changed in artificial ways.

The philosophy of natural behaviour seems then to be quite closely tied to the idea of a genetically coded nature, which in its turn predisposes to certain behaviour. If human beings interfere with this behaviour then they also inter-fere with the welfare of the animals. But is there any strict determination of behaviour? Certainly, there are behaviour patterns that are typical for certain species. Badgers and rabbits build burrows in their particular ways; birds of various species build nests in their particular ways, and have their particular ways of flying and singing. Some of these behaviours can be interfered with and this can create frustration.

But in many respects there is little determination. Birds of the same species typically choose different geographical sites for their nesting; they can live thousands of miles from each other. They can have variations in their food intake. Some chaffinches eat corn but some eat insects, and the same individ-ual can change its menu over time. And almost all animals every day introduce variations into their lives. They vary their movements and choose to see new locations. Like humans, many animals have a lot of real *freedom*. There is not much that is predetermined in how they live their lives in detail. What is natural is, given these facts, the freedom of the animals, or rather the potentiality for freedom.

The way humans can disturb animals should then be described as restric-tion of the free behaviour of animals. An animal can be prevented from build-ing a nest on the spot it has chosen. Animals can be caught and taken to zoos and hens and turkeys can be kept in small cages awaiting slaughter, measures that normally cause suffering for the animals.

It must be kept in mind, however, that it is not only humans that can inter-fere with the freedom of particular animals. There is interference within the animal world between the animals themselves, even within the same species. The predators continuously interfere with the predated animals. Such interfer-ence is continuous also in the sense that the predated animals must always be aware of dangers and behave cautiously even when the predators happen to be absent. Moreover, natural forces, such as earthquakes and sandstorms, can interfere with the intentions and behaviours of animals. All this happens in a 'natural' environment. There is a great lack of freedom also in the natural envi-ronment. But the human being is supposed to introduce an 'unnatural' restric-tion into the lives of animals.

We can ask whether the wild and natural life is normally a happy life to an animal. This is a relevant question since the naturalness is often assumed to bring happiness to the animal. The answer to this question must be qualified. Some animals, that are lucky and have few natural enemies, will probably live a reasonable life. Others clearly live a horrid and in many cases extremely short life. Nature has no mercy and the notion of help in nature is almost non-exis-tent. Help, on the other hand, is a notion in the human world that can become relevant to such animals as enter the circle of humans. Livestock, zoo animals and pets are fed, taken care of and frequently provided with medical care by their human owners. (Alrøe *et al.* concede this point in the following: 'On the other hand, there may be a contradiction between welfare conceived as a natural life and as the sum of positive and negative feelings, insofar as a life rich in the expression of natural behavior does not necessarily result in greater sum of hedonistic welfare', 2001, p. 291.)

So why is the natural life suggested by the associations for organic farming as the paradigm life for animals? The answer must be that there is a focus on our treatment of much livestock. Some of these animals, in particular hens, have been kept in extremely restricted ways, for instance in small cages, and have been nurtured in a way that has only had the human goals in mind. Here the human notion of help has often been non-existent.

The naturalness that the organic farmers call for deals with such minimal requirements as letting the animals move about freely and have a slightly richer life than they are normally provided with. Fulfilling these requirements hardly presupposes a philosophy of natural behaviour. Taking the animal's point of view (prevention of suffering and promotion of happiness) could be a sufficient ideal.

This point brings me to a preliminary assessment of the idea of natural behaviour as a candidate for a definition of welfare. It is questionable whether natural behaviour as such (i.e. behaviour not interfered with by humans) can *define* welfare. If this were the case, then behaviour aborted by an earthquake or by a leopard would constitute welfare. This is not what is intended. The root of the philosophy of natural behaviour must be that an animal flourishes and is happy when it has the freedom to fulfil its potential. But then the definition of welfare is not the natural behaviour as such; the definition turns on the notion of freedom of behaviour, or the freedom to express one's interests to use an expression in Rollin's philosophy (1992, p. 76).

The question now arises: is freedom of behaviour an intrinsic good? Or is freedom good because it is often conducive to happiness? On this point theorists differ. All the utilitarian ethicists would say that freedom is only an instrument. But animal theorists such as Fraser *et al.* (1997) would claim that freedom or autonomy is a good in itself. Nussbaum (2004), who argues for a capability theory (developed by Sen; see for instance 1993), declares that capability is an intrinsic value in itself. I am myself much leaning towards a happiness theory. However, as I will argue in Chapter 21, there are cases (for instance when we deal with non-sentient beings) where capability and freedom can in themselves be sufficient criteria for flourishing and welfare.

Bo Algers's Idea of Natural Behaviour

Bo Algers has in a number of articles (for instance Algers and Jensen, 1985; Algers, 1990) also introduced the notion of natural behaviour in contexts of welfare. (He does not, however, attempt a full definition of welfare.) In 1990 he took as a starting-point the current use of the term 'natural behaviour' in government texts and guides of various kinds. He noted, for instance, that the

Swedish government bill 1987/88:93 has the following formulation: 'There are methods of production that do not take the natural behaviour of animals into consideration.' And the implicit sense here is obvious. Such methods of production are unethical because they do not care for the welfare of the animals in question.

In his analysis of the notion of natural behaviour Algers emphasizes that natural behaviour is not simply the behaviour that animals exhibit in nature as opposed to situations of breeding. Such a simplistic interpretation would exclude dogs and cows from exhibiting any natural behaviour at all. Nor is natural behaviour the most frequent behaviour that is exhibited by an animal. Under such an interpretation a behaviour that happens to be statistically abnormal would automatically be counted as unnatural. But one could imagine very untypical behaviour, elicited in an untypical situation, for instance a situation of danger, that we would be inclined to call natural.

Algers finds that a better proposal for a definition is that natural behaviour is the behaviour that is normal for an animal in its normal biotope given the stimuli that are occurring. Algers explicates this notion by introducing the notions of *is-value* and *ought-value*. (See Wiepkema, 1985, 1987; Wiepkema *et al.*, 1993.) The is-value is identical with the information that the animal gets through perception. The ought-value is the goal of the animal (which is determined by the genes and the on-going evolution of the animal). Algers seems to assume that this goal and the gene programme associated with it is ultimately geared to the survival and reproduction of the individual. Thus this notion is tied to the notion of fitness as explicated above.

As long as the animal can use the is-value and reach (or at least approach) the ought-value, then the animal expresses natural behaviour. When this becomes impossible, in particular when the animal is prevented from moving from the is-value to the ought-value, then the animal often exhibits unnatural behaviour, i.e. behaviour that does not serve any purpose. One of Algers's examples is that of a cow that is locked in a small stall that is such that it is difficult or impossible for the cow to stand up. There is not enough space for the cow to move its centre of gravity forward enough to be able to rise. As a result the cow may exhibit frustration in various ways: emit noises, shake its head or perform other dysfunctional behaviour. If we talk in terms of evolutionary theory (which Algers does not explicitly do) then the natural behaviour can be equated with *adaptive* behaviour.

In the article quoted (1990) Algers does not explicitly define welfare in terms of natural behaviour. It is reasonable, however, to make the connection to welfare given the prevalent use of the term 'natural behaviour'. Consider, for instance, the above quotation from the Swedish government bill.

Assume that we say that an animal that in general exhibits natural behaviour has a high degree of welfare. This is obviously a characterization that is quite close to Broom's idea of animal welfare. In his terminology the same situation would be described as an animal *coping* with its surroundings. Natural behaviour contributes to fitness; the same holds for coping. In Algers's case, however, coping is limited to behaviour and does not cover the physiological function of organs.

Thus, if Algers's notion of natural behaviour is equated with Broom's coping behaviour, then the same arguments can be raised with respect to the idea that welfare is natural behaviour (or that welfare is the case when natural behaviour is the case) as used above with respect to coping. An animal can cope and can exhibit natural behaviour even in a difficult and harsh situation. But are we inclined to say that the welfare of the animal is perfect in all such situations? A strong animal can suffer, yet cope and exhibit natural behaviour. But many of us would still say that such an animal does not have a high degree of welfare. (Consider my distinction between coping as a process and coping as a final state concept in Chapter 9.)

In the final analysis Algers's and Rollin's conceptions of natural behaviour turn out to be quite similar. In neither case natural behaviour is identical with typical or common behaviour. Nor is natural behaviour necessarily the same as behaviour unrestricted by human intervention. What is in focus in both views is the animal's capability and freedom to realize its goals. At most there may be a difference in the interpretation of the notion of a goal. Rollin refers to the interests of the animal. Algers emphasizes the ought-value determined by the genes of the animal.

17 On Complex Views of Animal Welfare

Welfare as an Evaluative Concept: David Fraser

In a number of highly significant papers David Fraser (for instance, Fraser, 1995, 1999; Fraser *et al.*, 1997) scrutinizes the various proposals for the characterization of welfare discussed above, primarily from the viewpoint of our ethical concerns. A major starting-point for his analysis is that the concept of welfare is an *evaluative* concept. Thus welfare cannot be measured in a straightforward way. There must first be an evaluative choice with regard to what should count as welfare. After that one may look into such empirically verifiable parameters as contribute to welfare as defined from an evaluative point of view.

Given this fundamental platform for his thinking Fraser wishes to establish a scientific model of welfare. This means that he wants to select such values as regards quality of animal life as can form the core of a welfare concept and thus guide what empirical criteria we should use in establishing the welfare of animals. The method for selecting the values, in his case, seems to be a kind of consensus. Fraser (1995) selects three kinds of values referring to wide-spread consensus. He says (p. 112) that there is a large degree of consensus that 'a high level of welfare implies freedom from suffering in the sense of intense and prolonged pain, fear, distress, discomfort, hunger and thirst'. It is also, he says, widely agreed that welfare requires a high level of biological functioning. Fraser also mentions that there is a great, although not universal, agreement that animals should have positive experiences such as comfort or contentment.

Fraser *et al.* (1997), working in the same spirit, use a more specific starting-point in their choice of values. A scientific conception of welfare, they claim, must be based on an appropriate relation to the major ethical concerns that have given rise to animal welfare research. So here Fraser and colleagues more clearly place their discussion in the context of animal ethics. The choice of values, then, becomes similar although not identical to the one presented in Fraser (1995): (i) welfare presupposes that animals lead a natural life; (ii) animals should feel well and be free from prolonged negative states; and (iii) animals should function well in terms of physiology and behavioural systems.

I will first comment on Fraser's contribution to the theory of concepts. Fraser (1995) distinguishes between three types of concepts: type 1, type 2 and type 3 concepts. His paradigm example to illustrate this distinction is the following. A building inspector has been asked to examine a dubious rooming-house and report to the local authorities on: (i) its height; (ii) the weight of sun-bathers that would make the balcony collapse; and (iii) the building's overall safety. The height of the building can be measured directly. Thus height, con-stituting a single attribute, is a type 1 concept. The weight of sunbathers cannot be measured through a single measurement. The estimation presup-poses a number of considerations and measurements of various parameters pertaining to the balcony. Such a multidimensional, but still empirical, concept is a type 2 concept. It is complex but it is similar to a type 1 concept in that we presuppose that there is a single empirically correct estimate of this weight.

In describing the type 3 concept Fraser expresses himself in the following way:

> The safety of a building … is not a single attribute but a concept encompassing numerous attributes, including the level of toxins in the air, the slipperiness of the floors, and the soundness of the fire escapes. We regard these as aspects of safety because they serve a common function; that is, they all contribute in one way or the other to the health and survival of users of the building.
>
> (Fraser, 1995)

We might try to create a fire safety index but we recognize, says Fraser, that there is no correct quantitative expression that represents the safety of the building 'in all respects and for all users' (1995, pp. 104–105).

Strictly speaking, it does not follow from this characterization that type 3 concepts are evaluative ones. One may still (given the description above) char-acterize safety as an empirical concept but then as a concept denoting a set of relations. One may say that safety for person A exists under such and such conditions and safety for person B exists under such and such other condi-tions. These relations may in principle be empirically described. Thus, it does not necessarily follow from the mentioned premises that safety is an evaluative notion.

It is easy, however, to add to Fraser's matrix a fourth type of concept which is truly evaluative. An evaluative concept is then characterized by the fact that one cannot, from the definition of the concept, deduce how to empir-ically measure the magnitude denoted by the concept. In order to be able to measure one must fill in the blank space in the concept, i.e. one must choose what empirical facts constitute the goodness or badness in the relevant area of research. Thus, if animal welfare is a type 4 concept, i.e. a truly evaluative one, then we must decide, for instance through consensus, that certain things such as natural living, feelings and normal function constitute the welfare of animals.

Let me, to be cautious, add that a simple act of decision need not be sufficient for true evaluation. There is still room for argument. A person's value hierarchy must be consistent in order to count. If we can show that a person lives a life indicating that certain things are valued highly by him or her, but the person at the same time denies it in the verbal expression of values, then we can accuse him or her of inconsistency and ask for the true evaluation. Similarly, one may find that a particular consensus list of values is inconsistent or contains superfluous items in the sense that certain values are included in or follow from other values. Thus a more limited set may be said to contain the 'basic values'. Both critical techniques are common in welfare discussions. At times I am using such techniques in this book.

Welfare as an Integrative Concept

Fraser *et al*. (1997) are concerned, on the basis of the mentioned conceptual considerations, to meet the various ethical concerns for animal welfare in their characterization of the notion of welfare. These concerns have already been expressed in the preceding chapters of this book. Fraser *et al*. note, in particular, the following three fundamental ethical concerns in the animal welfare literature. 'First are natural living concerns which emphasize the naturalness of the circumstances in which animals are kept and the ability of an animal to live according to its "nature"' (p. 190). A second type of concern emphasizes the affective experiences of animals. The good life of the animal is thought to depend on freedom from suffering in the sense of pain, fear, hunger and other negative feeling states. 'Thirdly, functioning-based concerns, held especially by many farmers, veterinarians and others with practical responsibility for animal care, accord special importance to health and the "normal" or "satisfactory" functioning of the animal's biological system' (p. 191).

Fraser *et al*. also note that some theorists in the field propose definitions of welfare that cover all three of these concerns. The famous five-freedom concept, quoted above, is such a concept. Some scientists include all three factors but order them logically. Gonyou (1993), for instance, suggests that the animal's feelings are the ultimate concern but satisfactory biological functioning is important because it tends to influence the animal's feelings. However, say Fraser *et al*., the majority of theorists have adopted more limited notions of animal welfare centred on one or the other ethical concern. This, according to the authors, limits the scope of animal welfare research and it does not, they say, reflect all the quality-of-life concerns of the public. This general statement is supported in the article by a discussion of some consequences of a unilateral approach to animal welfare.

The idea that an animal's welfare depends solely on its being allowed to perform its natural behaviour is first scrutinized. This conception is for the most part compatible with feelings- and functioning-based interpretations, but it is sometimes more inclusive. Concern with regard to a bird's opportunity to fly is covered by the view that animals should be allowed to live according to their natural adaptations. But it is not so obviously covered by an emphasis on health, survival and subjective feelings.

On the other hand, as I have noted above and as Dawkins (1980) has elaborated, living a natural life is no guarantee that all ethical concerns will be satisfied. Even if an animal is kept in a natural environment it can still suffer and become ill. Wild animals are continually exposed to hazards such as disease and predation and indeed, as a result, captive animals may often be healthier and in this sense better off than wild ones.

Dawkins provides the following examples. The average adult song bird such as a robin or song sparrow, which might live for 11 years or so in captivity, lives only one or two years in the wild. Moreover the method of death may also cause suffering. Spotted hyenas, for instance do not have a specific killing bite or method of quickly killing their prey. As a result there may be many hours of terrible agony for the prey before it dies (Dawkins, 1980, p. 52). And how should one, say Fraser *et al.*, use the criterion of natural living 'in order to guide welfare issues concerning analgesia, euthanasia and medication' (1997, p. 193)?

Most writers on animal welfare sustain the idea that feelings play a part in welfare. However, they do this in different ways and Fraser *et al.* distinguish between three interpretations:

1. The term 'welfare' should be used only where feelings are involved.
2. Concerns about animal welfare arise because of the capacity of animals for subjective experience.
3. Concerns about animal welfare are concerns about the experience of animals.

Fraser *et al.* interpret 1 as a semantic claim that the term welfare should be applied only to situations involving subjective experiences. They consider that such a use could lead to confusion. 'According to established usage, welfare and well-being refer more broadly to "good fortune", health, happiness and prosperity, the state of being or doing well, thriving or successful progress in life, and a good or satisfactory condition of existence' (1997, p. 194). In particular, health has normally been included in welfare. Thus, a strict 'feelings-based definition of welfare might lead us to say that cigarettes reduce the health of smokers but may improve their welfare' (1997, p. 194).

Interpretation 2 only says that welfare is a concern for such living beings as can have subjective experiences. As far as we know only animals can have feelings. Thus only animals can have welfare.

Interpretation 3 says that concerns about an animal's quality of life are concerns about its subjective experiences. Changes in physiology, behaviour or living are of concern for welfare only insofar as they affect how the animal feels. This conception reflects the position of utilitarianism, Fraser *et al.* say: 'But such a restricted conception of animal welfare does not appear (initially at least) to cover the broader ethical concerns described above, such as protecting the health of animals and allowing them to live and develop according to their natures' (1997, p. 195).

In commenting on this position Fraser *et al.* note that even if one were to adopt the utilitarian stance, it does not follow that all animal welfare research should be limited to the subjective experiences of animals. The reason is that many ethically important questions about subjective experience cannot be answered empirically. We might, for example, ask whether keeping a bird in a cage reduces its welfare by depriving it of the pleasure of flying. But this question cannot be answered by empirical science: there is no acceptable method to quantify the pleasure experienced by the animal.

Disease and other disturbances of normal biological functioning are often the concern of veterinarians. Thus some theorists would characterize welfare in terms of normal functioning. Fraser *et al.* distinguish between three variants of this position.

1. All that matters is that animals function well.
2. We must study functioning because we cannot study feelings.
3. We can study functioning instead of feelings because the two are intimately related.

A proponent of position 1 is McGlone (1993), who proposes that an animal's welfare can be considered poor only when physiological systems are disturbed to the point that survival or reproduction is impaired. Fraser *et al.* reply to this, however, that if we limit our concern to physiological systems then we do not respond to the public's concerns concerning animals' subjective experiences, which are commonly seen as being of moral relevance in themselves.

Position 2 represents the positivist stance, according to Fraser *et al.* According to positivism science can only concern itself with observable phenomena. Feelings are not such phenomena. However, as I have argued in my general discussion of mental phenomena, most scientists have today distanced themselves from extreme positivism. So do Fraser *et al.* when they say that attempts to use the methods of science to improve our understanding of the subjective experiences of animals cannot be dismissed simply because such attempts depart from the influential positivist philosophy (1997, p. 198).

Position 3 represents the assumption that the functioning of animals completely reflects their subjective feelings. Thus the proponents of this position advocate studies of biological functioning as less problematic than such studies as attempt to measure feelings directly. However, there are salient cases of suffering in animals, Fraser *et al.* say, where there are no apparent changes in physiological functioning. Examples of such cases are when an animal has a strong inherited motivation for a certain behaviour, such as the sucking of a calf in order to drink. Today the calf can get water from a bucket and thereby satisfy its thirst. However, it still has a desire to suck that is not fulfilled. Hence, it suffers although this does not give any obvious physiological repercussions. On the other hand, sometimes functioning can be impaired because of the presence of pathogens. These changes need not be observed by the animal until at a very late stage of a disease. In short, there is no simple one-to-one correspondence between normal functioning and the subjective experience of animals.

Fraser *et al*. finally collect their views in an integrative model that provides an alternative to a one-sided definition of animal welfare (Fig. 17.1). The model is *not*, however, formulated as an integrative definition. It is instead a model covering what the authors take to be the broad classes of problems that may arise when the powers of adaptation possessed by the animal imperfectly fit the challenges it faces in the circumstances in which it is kept. The model is designed as a Venn diagram with three areas illustrating the various forms of welfare problems.

Comments on Fraser's Analysis

The paper by Fraser *et al*. (1997) is instructive and helpful for the student of animal welfare. It gives a superb overview of present standpoints and it contains crucial insights in animal science. There are a few things to observe

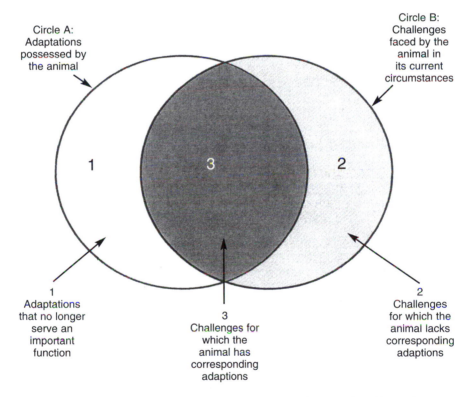

Fig. 17.1. Conceptual model illustrating three broad classes of problems that may arise when the adaptations possessed by the animal (circle A) make an imperfect fit to the challenges it faces in the circumstances in which it is kept (circle B). The authors propose that the different classes of problems, taken together, cover the major ethical concerns that arise over the quality of life of animals and constitute the subject matter of animal welfare research (from Fraser *et al*., 1997, with permission from the Universities Federation of Animal Welfare).

in their presentation and some of these observations may warrant criticism.

The authors discern three major positions with regard to defining welfare: in terms of natural living, in terms of feelings and in terms of normal biological functioning. On the surface of it all three positions are criticized for being too narrow. No single definition can respond to all the ethical concerns of the public. These concerns seem to function as the criteria of a correct definition of welfare. One can wonder whether Fraser *et al.* succeed in refuting all three kinds of definitions of welfare through their analysis. Consider in close detail the arguments.

Why is the natural living concept not functional? Fraser *et al.* say: 'Living in a natural manner is no guarantee that the full range of ethical concerns over the quality of life of animals will be satisfied. Even if an animal is kept in a natural environment where it can live according to its adaptations, it may still suffer and become ill' (1997, p. 193). Thus animals can suffer and become ill also in a natural setting and this is an ethical concern. Natural living thereby does not by itself cover the kernel of welfare. Instead natural living may be a means, although not a sufficient means, to attain welfare and welfare at least includes well-being (the opposite of suffering) and health.

The idea that feelings should be included in the definition of welfare therefore seems promising. But, Fraser *et al.* ask, should scientists restrict their conception of welfare only to the affective experience of animals? Feelings or experience are evidently a part of welfare, but do they exhaust the area of welfare?

The authors react to a definition solely based on feelings. Here they refer to current usage, according to which 'welfare' and 'well-being' denote a much broader set of factors including health, happiness and the state of being or doing well. Their crucial example concerns health. 'A strictly feelings-based definition of welfare might lead us to say that cigarettes reduce the health of smokers but may improve their welfare' (Fraser *et al.*, 1997, p. 194).

The authors also react to a rather similar proposal, viz. that concerns about animal welfare are concerns about the subjective experience of animals. This proposal is rejected by the authors. Animal welfare ethics/science must in practice deal with other things than subjective experience. This reason is methodological and empirical: there is not yet any accepted method to quantify and measure the pleasures and sufferings of animals.

This strategy of referring to methodological and practical difficulties cannot be accepted as a refutation of a feelings-based definition. The fact that some studies of animal welfare need to have something else than suffering as their immediate focus is no argument against a feelings-based definition. (Observe that virtually the same argument crops up from the point of view of the functioning definition. Here the authors, however, respond in a different way. They accept that it is possible to study experience in a scientific way.) The remaining forceful argument would be that *health* is an ethical concern and health is not solely defined in terms of feelings.

The authors finally tackle the normal functioning definition of welfare. This is also obviously too limited a definition, they say. 'If animal welfare research were to be limited in this manner, it obviously would not respond to the widely held ethical concerns over animal's subjective experiences' (Fraser *et al.*, 1997, p. 197). Here the criterion used is that the definition does not fit with the concern for suffering. Thus, suffering again comes up as a crucial element in welfare.

The result of Fraser's *et al.* analysis thus boils down to the following. Fraser *et al.* wish to show that all the three major attempts to define welfare are equally one-sided; they fail to take account of all ethical concerns. However, if one scrutinizes the arguments it seems not to be the case that they are equally limited. To put it simply: natural living is inadequate because it does not preclude suffering and illness. Natural functioning is limited because it does not preclude suffering and suffering is accepted as a central ethical concern. Thus Fraser *et al.* in practice come much closer to the feelings definition of welfare than to any of the other ones. The major problem with the feelings definition is that it does not exhaust what welfare, according to established usage, denotes. But this is a problem shared with the other definitions. The major factor which must be included among the welfare values, according to Fraser *et al.*, and which is not included in a feelings definition, is health.

A crucial observation is the following. On one level of discourse all three ethical concerns are equally important. Farmers and pet owners and other people involved are concerned about the natural living and the feelings of animals, as well as their natural function. However, a closer analysis may end up in the conclusion that there are strong dependencies between these values, as I was indicating in the last section. The value of natural living may be (at least partly) dependent on its relation to the long-term subjective comfort of the animal in question. The value of natural function may be (at least partly) dependent on its relation to the long-term subjective well-being of the animal. But there is no symmetry here. The value of positive feelings is rarely dependent on its relation to natural living or natural function. I admit that one can envisage cases of such dependence. The positive feelings induced by drugs and alcohol are not valued greatly, since they are dependent on unnatural or even detrimental physiological function. On the other hand such positive feelings are not enduring feelings. Long-term positive feelings seem to retain their high value irrespective of natural living or natural function.

This is not an argument for saying that feelings are the only welfare values that there are. In my own conclusion (which attempts to incorporate also non-vertebrates) I will argue for a more inclusive concept of welfare. I wish, however, to emphasize that feelings (and, basically, preferences) have a very strong position among our welfare values.

The Idea of Complex Definitions under Scrutiny

Other authors side with Fraser *et al.* (1997) in their belief that one cannot propose a simple theory of human welfare. Alrøe *et al.* (2001) claim in their work that a systemic view, which is central to the idea of organic farming, calls for a more complex notion of welfare than most theorists propose. We cannot simply consider fulfilment of an animal's basic needs. We must also consider such factors as the richness of the animal's life and consider its relation to the whole ecosystem. 'Organic farming's declared holistic ethos means that the solution of welfare problems, with their concomitant ethical considerations, must necessarily be discussed from a very broad perspective that takes the whole agricultural system into consideration' (p. 292).

A similar broad approach to animal welfare (or 'well-being', which is the term they prefer) is adopted by J.D. Clark *et al.* (1997), who make the following policy statement:

> We believe that well-being is multifaceted, and factors that affect it are interactive and interrelated. An animal's well-being or quality of life is its internal somatic and mental state that is affected by what it knows (cognition) or perceives, its feelings (affect) and motivational state, and the responses to internal and external stimuli or environments.
>
> (Clark *et al.*, 1997, p. 566).

This brings me to make a deeper analysis of the idea of a broad conception of animal welfare and see to what extent it could be useful. Consider the following interpretations of this idea:

1. *No differentiation made between concepts, criteria and indicators.* In some conceptual analyses a number of features are mentioned which are said to have a relation to the concept in question. In the case of welfare it can be a question of features such as coping function, naturalness, feelings, preferences, illness, suffering, abnormal behaviour, stress, hormones and self-narcotization. The relations between these features and the concept of welfare are obviously different but this is not always made clear. Sometimes, however, the impression is given that all features are in one way or the other components of the concept of welfare. This is what I wish to call a *superficial* (and for most purposes gravely misleading) idea of a broad conception of welfare.

In fact, given any reasonable account of welfare, some of these features, for instance feelings or coping function, are parts of the concept of welfare – and should be mentioned in an adequate definition of the concept – whereas others, for instance the non-existence of diseases, may be criteria of welfare. Yet other features are more or less reliable indicators of the existence of welfare.

2. *Complex notions of welfare.* A theorist who is aware of the distinctions made above can still claim that there are and should be broad concepts of welfare. And the idea, following Fraser *et al.* and Alrøe *et al.*, could be that the definition of welfare requires the inclusion of at least the following three components: function, feelings and natural behaviour. Thus the determination of an animal's welfare must be dependent on the assessment of features on all three scales.

Consider now three different concepts of complexity:

1. *Specificity.* 'Car' is a more complex concept than 'vehicle' in the sense that a car must exhibit more characteristics in order to be a car. A vehicle is a movable object that has any kind of supports, whereas a car is a motorized vehicle running on wheels. In order to assess whether an object is a car we must therefore verify the existence of all three characteristics, viz. the vehicle, an engine and the wheels.

2. *Split concepts.* Another idea of complexity of concepts is more spurious. The idea is essentially that a term 'X' can stand for different concepts. In essence, 'X' is ambiguous as between the concepts. Example: 'bank' can mean either 'financial institution' or 'heap of sand'. This salient ambiguity is rarely referred to, however, as 'complex concepts'. A situation referred to as 'complex concepts' can arise, however, when the concepts involved stem from the same area of research and when there is controversy about their definition. This is the case with welfare.

In the case of split concepts one can formulate definitions in two ways. One way is to clearly demarcate various senses of, for instance, the term 'welfare' as welfare in sense 1, welfare in sense 2, etc. In both senses we can talk about high degrees of welfare, even optimal welfare. A dog can have optimal welfare according to sense 1 but less than optimal welfare according to sense 2. Another technique is to make welfare a disjunctive concept: by 'welfare' is meant *either* 'coping function' *or* 'positive feelings'.

3. *Conglomerate concepts.* A third sense of complexity is one where a number of more simple concepts have been put together to form a new, complex concept. Welfare, according to this conception, is a conglomerate of, for instance, coping function, feelings and naturalness in behaviour. It is not either one or other of these features.

A consequence of this interpretation is that perfect coping function is not sufficient for perfect welfare. Presumably, the animal must reach perfect states also along the other dimensions in order to reach perfect welfare. The question can also arise whether perfect welfare can occur at all given this interpretation. It has been observed that at least some instances of a high degree of welfare on the natural behaviour dimension may conflict with a high degree of welfare in the feeling sense. Alrøe *et al.* (2001) themselves make the observation: 'There may be a contradiction between welfare conceived as a natural life and as the sum of positive and negative feelings, insofar as a life rich in the expression of natural behaviour does not necessarily result in a greater sum of hedonistic welfare' (p. 291).

What is or what could be the sense intended by the locution 'a broad concept of welfare'? If welfare were to be complex in the first sense then it would be a species of a genus. What could be the candidate for being such a genus? Health might be a candidate. A person in good health must fulfil certain conditions, e.g. must lack all disease. An animal having welfare (a species of health) must first exhibit the basic conditions of health but also have a certain subjective well-being implying that it is not suffering. This is indeed a reasonable interpretation and some authors have come close to it.

Do people say that welfare is *ambiguous*, i.e. that there are different welfare concepts? This is indeed commonly said in certain review papers, such as Duncan and Fraser's (1997), where they present some of the concepts proposed in the animal science literature. The feeling concept claims universality in the sense that the animal that feels perfectly well has perfect welfare. And the same holds *mutatis mutandis* for the other concepts. It is rarer, though, that a particular theorist claims that there are three *equally valid* welfare concepts. The theorists typically choose and argue for one concept. See what Broom, Dawkins, McGlone and Duncan say in their own theoretical writings.

The most plausible interpretation of the complex definition claim is to say that welfare is a *conglomerate* concept. Welfare is not a species of anything else, for instance health; welfare is rather the sum total of states along a number of dimensions, for instance coping function, feelings and natural behaviour. In order to have a high degree of welfare one must score high on all dimensions. It is also natural to interpret Fraser *et al.* (1997) (see above) as providing a conglomerate concept of welfare.

Questions concerning the balance between the dimensions, for instance whether a high score on one dimension in some way can compensate for a lower score on another, have to be settled by a primary evaluation made by the assessors. There is nothing here that follows from the conceptual analysis in itself. The assessors must judge whether any one of the three dimensions is more important than any other dimension. This means that when a complex concept of this kind is put to use there will appear a number of difficult choices for the operationalization of the concept, in this case for the measurement of welfare. Alrøe *et al.* are aware of this problem and are prepared to face it.

My own strategy, to be pursued below, is to show that we may not need a conglomerate concept to express what we wish to express in this area. The complexity that is superficially seen can often be resolved by distinguishing between concepts such as external and internal welfare, between coping and the experiences involved in coping, between elements that are logically relevant and such as are causally or otherwise empirically relevant to each other.

18 On Conflicts Between Individual and Systemic Welfare

In addition to the broad concept of animal welfare the ideology of organic farming introduces another complexity into the theoretical arena. Its proponents do not only wish to discuss individual welfare but also and indeed primarily *systemic welfare*. Alrøe *et al*. say:

> Behind the objectives lies the fundamental tenet of the organic movement – that humankind is an integral part of nature … This systemic conception of agriculture that emphasizes the interaction between humans and nature is fundamental to an examination of animal welfare in organic farming. Livestock are an element of this interaction, and often an important one.
>
> (Alrøe *et al*., 2001, p. 280)

This systemic hypothesis leads, according to the organic farming proponents, to a systemic ethics. 'Organic farming's declared holistic ethos means that the solution of welfare problems, with their concomitant ethical considerations, must necessarily be discussed from a very broad perspective that takes the whole agricultural system into consideration' (Alrøe *et al*., 2001, p. 292). Thus we are not only talking about the welfare of individual animals but also about the welfare of the whole ecosystem.

An obvious consequence of this conception is that conflicts may arise between the welfare of the individual and the systemic welfare. This is noted by Alrøe *et al*. (2001):

> The above-mentioned EU directive also emphasizes the use of rearing and farming methods that promote resistibility and strengthen the animal's natural immune defence. Moreover, it states a number of restrictions on the use of medicine, including increased holdback times and requirements for renewed conversion in case of repeated treatments. These restrictions are not based on an individualistic view of animal health and welfare. On the contrary, they can

lead to inferior welfare for the individual animal, because of insufficient
treatment. Such rules can only be understood from a systemic view of the
solution of welfare problems, comparable to restrictions on use of pesticides and
artificial fertilizers in organic plant production.

(Alrøe et al., 2001, p. 294)

For instance, withholding the use of medicine in treating a particular animal,
which may lower the welfare of this animal, could improve the welfare of the
whole system. Thus there is a conflict in this case between the enhancement
of individual welfare and the enhancement of systemic welfare.

Here there are a number of issues that have to be disentangled. One question, not answered in Alrøe et al.'s analysis, is: what constitutes systemic
welfare? First, how is the ontology of the system to be described? Is the system
the sum total of individual plants and animals, including humans? Or is the
system something over and above the sum total of individuals? What is the
concept of welfare used in the case of systems? Is it a broad concept analogous to the one for individuals?

Assume first that the system is constituted by the sum total of individuals
and plants in the part of the world analysed. We can then easily apply the
same concept of welfare to all the individuals (for the case of plants, however,
the feeling component has to be excluded). It is then no wonder that the
welfare of one individual may come into conflict with the welfare of another
individual or even most other individuals. This is one of the most common conflicts in human ethics and it can often be resolved by considering the number
of individuals involved and the strength and duration of the pleasure/suffering
involved. A brief period of suffering on the part of one individual must stand
back for long-term suffering on the part of a whole population. So it can be
argued, on ordinary ethical premises, that if the abstention from medication on
the part of one individual may have strong positive repercussions on the
immune defence systems of a whole population, then the ethical choice is
easy.

This, however, need not be expressed as a conflict between two kinds of
ethics, one individual and one systemic. There is one ethics taking care of a
whole population of individuals.

Assume now – which is the most reasonable interpretation – that the
system is interpreted as something over and above the sum of the individual
animals and plants. What could the system consist of over and above animals
and plants? There are, for example, meteorological conditions, i.e. there is a
climate; there are numerous relations between the climate and nature, as well
as between elements of nature; there is growth, development, decay and death
in the system. Assume that we wish to talk about the welfare of the sum total
of all these factors: how should this be interpreted? Can we sensibly talk about
the welfare of any of them? What is the welfare of the climate? What is the
welfare of the relations between the parts of the system? What is the welfare
of the development of the system?

It is obvious that we cannot ascribe a concept of welfare taken from the
human or the animal world to an entity such as the climate. The climate is not
a biological being. So, welfare here must be something completely different.

Nor can welfare in the above sense be ascribed to the relations (intentional and causal relations, for instance). The relations are not sentient and biological beings. Perhaps in an extended sense welfare can be ascribed to the development of nature. The growth and development of the system can be measured in quantitative terms. On the other hand, can we say that the growth of the system *ipso facto* constitutes its welfare? Does the system have welfare irrespective of the welfare of the animal and plant elements of the system?

A plausible proposal could be that the welfare of the system as a whole is its fitness, i.e. its capability for survival as a system. We might then use a *part of* the complex concept of animal welfare. A particular part of nature is more or less fit. It can have resistance to various threats; for instance the rainforest in Brazil can have welfare, not in the sense that it has any feelings but in the sense that it can cope with meteorological threats and the threats imposed by human industrialization and urbanization.

This may be a completely alternative analysis but it need not be. The crucial issue is whether the welfare of the system is determined for its own sake or in relation to humans and animals. We may think that the value of the rainforest in Brazil ultimately rests on its importance for human and animal survival on Earth. Then, again, the welfare of the system boils down to its causal strength in supporting animal and human welfare. The system interpretation of welfare is therefore an alternative analysis only insofar as we evaluate the survival of the system *for its own sake*, not at all with regard to its importance for human or animal survival and welfare. I doubt, however, that there is a serious proponent of a systems analysis that is unconnected to individual human and animal welfare.

The only reasonable way, it seems to me, to include the climate, various abstract relations, and also growth and development, in a discussion of welfare is to see how all these elements and interactions *contribute* to the welfare of the majority of plants, animals and humans. Nobody could deny that there is a huge systemic interaction in nature and that this interaction is an extremely complex set of conditions for the welfare of all the various individuals. This means that, when we consider interventions in nature, for the sake of the welfare of a certain individual or for most individuals, we must carefully analyse this complex web of interactions. (This fact is well discussed and illustrated in Lund and Röcklingsberg, 2001. However, in contradistinction to my approach they actually propose different kinds of welfare for the different systemic levels.)

But if my analysis holds true, then the welfare of the system boils down to the welfare of the individuals in the system. But one should of course add the important factor of time to the analysis. We should be concerned about the *sustainable* welfare of individual humans and animals.

19 Welfare and Time

Except in the case of a few authors, for instance Broom and Johnson (1993), Simonsen (1996) and Broom (1998), the time factor in the assessment of welfare has been notoriously neglected. An animal can have welfare for a short time and for a long time and we can work for the short-term or the long-term welfare of the animal. This elementary observation plays a great role in the quality-of-life debate with regard to humans. The York school of health economists have introduced the notion of a 'qualy', which is a unit for measuring quality of life in humans, and where the time factor is captured. For instance, to say that a surgical treatment results in ten 'qualys' means that a human being can live 10 years completely healthy after the treatment. (See Williams, 1995; Brooks, 1996.)

Some of the alleged conflicts noted in the animal science literature with regard to different assessment scales can be more easily analysed by introducing a time factor. There is some short-term pain (or other suffering) in all daily human and animal affairs. This pain or suffering is in fact often a necessary condition for more long-term satisfaction and well-being. Searching for food can be both tiring and scary for a wild animal. The animal may have to struggle and even be slightly injured in order to succeed in its endeavour. But we would not usually say that these sensational signals constituted reduced welfare for this animal. Some of our desires, for instance the desire to indulge in sweets, both in the human and the animal world, may have to be frustrated and thereby create suffering, in order for more long-term pleasure to be obtained. The vital event of giving birth can entail considerable, sometimes unbearable, suffering to the mother, without us talking about a reduction in her quality of life. The suffering is (in standard cases) terminated within a few hours and it is a precondition for life-long happiness.

The time factor also comes in when one analyses the potential conflict between a hedonistic approach to animal welfare and a preference–

satisfaction approach. Several theorists, including Broom and Alröe *et al.*, have noted that an assessment of welfare based on suffering can sometimes differ from an assessment based on preferences. The obvious reason is – and this is indeed also common among humans – that the individual may not know what is best for it, and prefer what looks good but what will in the long run be detrimental and create suffering. The easy example in the human case is the individual's desire for sweets. Sweets may be the object of extreme desire and thus the satisfaction of this desire comes high on a preference–satisfaction scale. On the other hand, on a scale that takes the consequences in the long run into account, the satisfaction of the desire will score badly.

The crucial thing here is that the two scales make different uses of time. Preference–satisfaction (in the simple form) only considers a short period. The suffering scale may consider a much longer period. It is easy, however, to use the idea of preference–satisfaction in a more sophisticated way. First, we should talk about *lasting* preferences. We should rely on preferences of the animal or the person who has considerable experience of making a variety of choices or performing a variety of activities. Thus the preference can become an *informed* preference. We know that not only humans but also many animals can be influenced by previous experiences in their choices. As a result they will choose a strategy that they believe to be satisfactory in the long run. Of course, different organisms have different abilities in making such predictions.

20 Summing up the Analysis

What are the conclusions to be drawn from this comparison between the notions of human and animal health and welfare?

There is a great difference between the two areas with regard to the notion of health (and the allied concepts illness and disease). There is very little scientific discussion of the concept of health in veterinary or animal science. Gunnarsson's analysis of veterinary textbooks and encyclopaedias shows that views exist on the nature of health, in terms of, for instance, balance, natural function and well-being, also in terms of production. However, with the exception of Broom's recent writings and Hovi *et al.* (2004), there exist quite few arguments for or against particular conceptions of health. It appears that most writers in veterinary and animal science tacitly presuppose a notion of health in terms of some natural function. This notion is, however, rarely analysed in the literature.

The situation is quite different in the case of humans. Since around 1970 there has been an advanced discussion on the nature of health within several disciplines, including anthropology, sociology and philosophy. A variety of standpoints have developed, from a naturalistic one to a psychosocial and normative one. The American Christopher Boorse initially (1977) presented a well-argued naturalistic theory. This theory has given rise to a multitude of responses and new positions. Boorse has given his answer to these responses in a long article (1997) and the debate goes on from there (Nordenfelt, 2001).

One can speculate about the reasons for this salient difference. One reason is obvious. Human health care is of concern to all. We worry about our own health and the health of our nearest and dearest. But the concern is also economic. The health-care budget comprises 10% of the total national budgets in the world. We need to know for what kind of health we spend all this money, and in particular to what level of health we can and should aspire in our health care and health promotion.

Although a concern indeed exists also for livestock and pets in this regard, the difference in relation to the human case is enormous. The health of animals is often seen to function as a means for the realization of other ends, such as the production of milk and meat, or our own entertainment. And although we must not dismiss the existence of genuine concern for individual animals, this concern cannot be comparable to that which exists for human beings and their individual health.

So why is health so crucial to humans? The answer is simple. Health is important for a person because health enables this person to engage in crucial activities, such as work, political activities and leisure activities, and not least, to engage in close human relations, such as friendship and love. The importance lies precisely on the psychosocial level. Thus the health concept that is of interest to people is a holistic concept embracing all the relevant abilities. It is not a strictly biological one. A person has little concern for his or her biology as long as the body works on the molar level. Here is one explanation of the fact that serious holistic health concepts have been proposed rivalling the purely biological ones.

With regard to welfare the picture is much more complex. Here we can find so many diverse interests, not only within the disciplines of human and veterinary medicine. In the human case the question of welfare has had a central position throughout the history of ideas. The great philosophers, such as Plato and Aristotle, have dedicated some of their most important works to the question of welfare or its close relation happiness. Welfare and happiness are the most crucial notions in consequentialist ethics, in particular utilitarian ethics. Thus these concepts require a central place in all ethical discussion. It is only rather late (i.e. in the late 20th century) that they have received a core position in the social services as well as in medicine and health care.

So why has this development come about? Let me only focus on health care and here the development is in line with the focus on a holistic concept of health. The general public has gradually acquired more interest in and influence on health care. They wish to see that health care aims at the crucial things, viz. holistic health and what is often called 'health-related quality of life'. The healthy person should be able to do relevant things but should also feel well and in general be content with life. Thus great interest has developed in finding tools for assessing and even measuring these phenomena.

The emergence of a welfare discussion in the animal field has another kind of source that I have noted in the Introduction. With regard to animals I observed that animals have in many situations been treated with negligence and even with cruelty. It is not only that some veterinarians or animal keepers have cared little for the individual animal unless the situation has had economic consequences. One can even find examples of horrible torture in the handling of animals. Thus a strong ethical concern has emerged for the *protection* of animals from human cruelty. Comparable events have occurred also in the history of human affairs. Salient examples can be found in the history of medical research on humans, in particular during the Nazi regime in Germany. Such situations are not, however, as far as we know, frequent today.

Given the different motives behind the discussions in the human and the animal field one could expect different emphases in the respective analyses of welfare. This indeed is to some extent the case. One can find a cluster of human concepts focusing on freedom and abilities, and one can find a cluster of animal concepts focusing on suffering and well-being. The difference is, however, less than one could have expected. Physical well-being is central in most human concepts of welfare or quality of life, and coping and natural behaviour are essential elements in some animal concepts of welfare. An explanation of this fact is probably that the analysis of welfare has to some extent retired from its ethical starting-point and become 'scienticized'. The study of welfare is a discipline on its own and has assumed the aim of making a general analysis that should be able to serve a multitude of purposes, not only in ethics.

I will now list some major theories of health and welfare with regard to animals and humans and finally make some observations with regard to their similarities and differences.

On human health
1. Balance and homeostasis theories (the oriental tradition, Galen and Cannon).
2. Health as natural function (the biostatistical theory of Boorse). This idea is often coupled with the thesis that health is the absence of disease.
3. Holistic theories I: well-being (Canguilhem).
4. Holistic theories II: ability and well-functioning (Fulford, Pörn, Nordenfelt).
5. Combined theories (between natural function and holism) (Wakefield).

On animal health
Not many theories but one can discern several different views in textbooks and encyclopaedias.

1. Homeostasis views (Day).
2. Health as biological and natural (normal) function (Cheville).
3. Health as well-being (Blood and Studdert).
4. Health as productivity (Grunsell).
5. Combined definitions (Blood and Studdert).

On human welfare or quality of life
1. *Eudaimonia* (virtuous activity) (Aristotle).
2. Pleasure (Bentham).
3. Quality of life = subjective well-being (Veenhoven).
4. Quality of life = fulfilled preferences (Nordenfelt).
5. Fulfilment of needs (Aggernaes and Moum).
6. Quality of life, a mixed concept containing objective properties (Brülde, Naess, Kajandi and several others).

On animal welfare
1. Development according to natural selection (Barnard and Hurst).
2. Coping (Broom).

3. Well-being (Dawkins and Duncan).
4. Satisfaction of preferences (Sandøe).
5. Fulfilment of needs (Hughes and Curtis).
6. Natural behaviour (Alrøe, Rollin and Algers).
7. A conglomerate notion of welfare (Fraser).

Concluding Observations

There are a great variety of proposals for the analysis of the concepts of health and welfare both in human and animal affairs. Let me here attempt to list the most salient similarities and differences.

Similarities with regard to health
1. In both areas we find theories of biological, natural or normal, function.
2. In both areas we find ideas of biological and psychological balance or homeostasis.
3. In both areas we find ideas focusing on well-being.

Differences with regard to health
1. Health is regarded as a controversial concept (almost) only in the human discussion.
2. In the human health discussion there are a multitude of psychosocial concepts of health, emphasizing the healthy person's ability to realize goals in society.
3. Theories relating health to production occur only in the animal discussion.

Similarities with regard to welfare
We find great similarities in the theories presented, for instance with regard to well-being, need-fulfilment and preference-satisfaction.

Differences with regard to welfare
1. Biological theories, dealing with coping, occur only in the animal field.
2. Theories on natural behaviour occur only in the animal discussion.

I will now turn to a more systematic approach to the theory of health and welfare, taking my departure from some theories with regard to the human case.

III A Holistic Approach to Animal Health and Welfare

21 Towards a Holistic Theory of Health in Animal Science

Introduction

Above I have introduced some views in human and animal science with regard to health and welfare. I have laid most of my emphasis on the animal science literature but I will now use arguments from the human field to try to reconstruct the conceptual framework also for animal science. What I seek to do here is to sketch a holistic, action-theoretic, analysis of health and disease also for the animal area. I will later propose a preference-based theory of animal welfare.

The starting-point for my analysis will be the biostatistical theory of disease developed by Boorse (1977, 1997): 'A *disease* is a type of internal state which is either an impairment of normal functional ability, i.e. a reduction of one or more functional abilities below typical efficiency, or a limitation on functional ability caused by environmental agents' and health is identical with the absence of disease.

The crucial concept here is functional ability. This has been explained at other places in Boorse's writings. An organ exercises its function, for instance the heart is pumping in the appropriate way, when it makes its species-typical contribution to the individual's survival and reproduction. Survival and reproduction are the crucial biological goals, according to Boorse, determining the notion of biological function.

I have earlier (1987/1995, 2001) criticized this analysis of health from several points of view. I will here present two arguments: first the argument from repair, second the argument from the clinic.

The Argument from Repair

In Boorse's first presentations of his theory he did not consider the importance of the environment for human functioning. All physiologists and biochemists know that, depending on the circumstances as well as the individual's food intake and movements, the functioning of the human body will vary. A person who is standing in the cold will be freezing, a person who has just eaten a lot of food will have a lot of intestinal activity and a person who has just been exercising will have an extreme increase of his or her heart and lung functioning. All these changes in the functioning of the human body are considered to be 'normal'. But they are statistically normal only if we make this concept a very complex one and calculate the statistically normal functioning given all conceivable circumstances.

Such a calculation is, however, in principle feasible, so this observation is not automatically a blow to the biostatistical theory as such. One can say that the functioning of the heart of the athlete who has just run 100 m (and whose pulse is 180) is normal given the particular circumstance of this athlete's 100-m race. It is more interesting, however, to extend the argument to really harsh circumstances. Consider the case of an ordinary infection. Microbes invade the organism. The immune defence system is triggered and starts its combating work, terminating in the annihilation of the microbes. This work of the immune system is 'statistically normal' given the particular circumstances of the microbe invasion. The snag is, however, that this 'statistically normal' work of the reparative system is what constitutes the major part of the illness, as experienced by the individual. It is the repairing work that causes the pain and the fever, not the microbe invasion as such. The paradoxical conclusion of this argument then is: the immune response to the infection makes its statistically normal contribution to the individual's survival. Thus this state of affairs is healthy. On the other hand, a situation of this kind, perhaps involving pain and high fever, is nevertheless a typical instance of human illness.

There are conceivable moves to be made here. Boorse has chosen one that I consider inadequate. His proposal is that the calculation of the normal functioning of the human body shall only take into account 'statistically normal circumstances', thus presumably excluding 'dangerous' circumstances. But how much is included in the span of statistically normal circumstances? We risk including too little. A 100-m race is hardly a statistically normal circumstance to most people. Yet it is hardly a dangerous circumstance. And we risk including too much. There are many types of circumstance that are extremely common in some parts of the world but which are still dangerous. I have been told that 90% of some African populations have bilharzia disease. The existence of the bilharzia worm is therefore statistically normal, at least in a great part of the world. So what should be the reference basis of the statistical calculation?

Another move is to say that in the case of the infection the 'real' disease precedes the repairing work of the immune system. The real disease would then be constituted of the 'damage' done in the cells invaded by the microbes. I do not doubt that there are cases of infection where there is great damage caused to central systems of the body in the sense that their function is reduced below typical efficiency before the start of the immune response. Therefore I do not doubt that the biostatistical approach will capture most instances of diseases as intuitively understood. I doubt, however, that it will capture *all* instances. Immunologists have told me that in some cases it is sufficient that microbes are present and start infusing their toxins for the immune system to start its work. Hardly any damage to the tissues needs to be present before the infection begins.

Making these observations one can wonder whether the statistical approach is the most reasonable approach to the philosophy of health and illness. If, in several cases, it is the statistically normal response of the organism to an external threat that causes the symptoms that we in ordinary discourse associate with illness, then one can wonder whether the conception of statistically normal functions is an adequate basis for a theory of disease and health.

The Argument from the Clinic

This brings me to the second kind of critical argument against the biostatistical conception. I call it the argument from the clinic and it is an argument that lies behind the Reverse theory of disease and illness (Canguilhem, 1978; Fulford, 1989; Nordenfelt, 2001). Consider the following, hopefully plausible, story with regard to the emergence of the concepts of illness and disease.

1. In the beginning there were people who experienced problems in and with themselves. They felt pain and fatigue and they found themselves unable to do what they could normally do. They experienced what we now call illnesses, which they located somewhere in their bodies and minds. Several people came to experience similar illnesses. This led to the giving of names to the illnesses, and hereby the presence of the illnesses could be efficiently communicated. These were the phases of *illness recognition* and *illness communication*.

2. The people who were ill approached experts, called doctors, in order to get help. They communicated their experiences to the doctors, via the illness language. The doctors tried to help them and cure them. In the search for curative remedies, the doctors did not just rely on the stories told by the people who were ill. They also looked for the *causes* of the illnesses within the bodies and minds of the ill. This meant in the end that they initiated systematic studies of the biology of their patients. This was the phase of *search for the causes of illnesses*.

3. As a result of these studies the doctors found some regular connections between certain bodily states and the illness symptoms of their patients. They formed hypotheses about causal connections between the internal states and processes and the illness syndromes. They designated these causes of illnesses *diseases* and they invented a vocabulary and a conceptual apparatus for the diseases. This was the phase of *disease recognition*.

4. Once the diseases were recognized, a new and independent research could be established. The diseases as biological processes could be studied in their own right and irrespective of their connections to human suffering. This was the phase of *biomedical research*.

The point of my story-telling is that in the beginning there is always a *problem* conceived by a subject. The problem is a vital problem, one that concerns the subject's living a reasonable life, but it does not necessarily concern a threat to the person's life, growth or reproduction. The problem quite often concerns pain or other kinds of suffering, such as depression or extreme fatigue. And the subject believes that this problem has some kind of internal (biological or psychological) cause. This is the standard condition of the patient who is seeking help in a health-care situation.

The focus of attention is thus the illness; the problem as perceived by the subject. I therefore consider illness to be the primary concept from the point of view of conceptual development. From the concept of illness we can derive the concept of disease, i.e. the internal state which causes (or tends to cause) the illness. But observe here how the diseases are identified. They are identified on the basis of an illness-recognition. A discovery of the disease presupposes the occurrence of an illness. (Hence the expression 'Reverse theory of disease and illness'.)

Given this interpretation we arrive at a definition of disease which is far from the Boorsian biostatistical one. A preliminary definition of disease would thus be: disease = a bodily or mental process which is such that it tends to cause an illness (understood as a state of suffering or disability in the subject).

The Central Place of Disability

Suffering and disability are not novel concepts in the history of medicine or health care. They are central in the clinic but they have not consistently been used in the theoretical characterization of basic medical concepts such as disease and pathology. In the history of philosophy of health, however, we can find the ideas of suffering and disability in the writings of certain authors. The *locus classicus* is Galen's formulation in his *Ars Medica* (*The Art of Medicine*) written in about AD 190. (For a modern edition, see Galen, 1997.) It runs: *Health is a state in which we neither suffer from evil nor are prevented from the functions of daily life*. Thus, health is here understood as the opposite of illness. (See also Temkin, 1963, p. 637.)

There are alternative ways of using these phenomena in the construction of theories. Either one uses both kinds of concepts, as do Galen and Reznek (1987), and say that illness is constituted both by suffering and by disability, or one focuses on one of them for the purpose of definition. In my own analysis of health I have focused on the concept pair ability and disability, since I find it to be more universally useful than the concept pair well-being and suffering. (Observe that this holds only for my analysis of health, not for my analysis of welfare or quality of life, where the mental concepts of well-being and suffering play a fundamental role.)

At the same time we realize that there is a strong connection between suffering and disability, where suffering is taken to be a highly general concept covering both physical pain and mental distress. A person cannot experience great suffering without evincing some degree of disability. But the converse relation does not always hold: a person may have a disability, and even be disabled in several respects, without suffering. There are, as I have noted above in Chapter 2, paradigm cases of ill-health where suffering is absent. As a consequence of this reasoning a number of theorists (Seedhouse, 1986; Nordenfelt 1987/1995; Fulford, 1989) have used ability and disability as the central notions in the definitions of health and illness.

An Analysis of Health in Terms of Ability to Realize Vital Goals

What, then, should a healthy person be able to do? Or, conversely, what kinds of disabilities constitute a reduction of the person's health? What disabilities are such that the health-care system should provide health care for the person? These questions are clearly not identical. The last question has a political overtone. The answer to it depends not only on conceptual analysis but equally on policy decisions concerning medical priorities. I shall not enter into a discussion of this here.

It is plausible to believe that whatever the adequate answer to the question of the nature of health should be, it will be an answer on an abstract level, capable of being summarized in terms of certain general goals. The question to be put should then rather be formulated in the following terms: what are the goals that a healthy person must be able to realize through his or her actions?

My general proposal (2001, p. 9) is the following: *A is completely healthy if, and only if, A is in a bodily and mental state which is such that A has the second-order ability to realize all his or her vital goals given a set of standard or otherwise reasonable conditions.* (I have here made a slight revision in relation to 2001.) Let me now clarify and to some extent defend this proposal by commenting on the crucial clauses concerning vital goals and second-order ability. I will be brief with regard to the first clause and instead concentrate on the relation between health and second-order ability.

What are the vital goals of a human being? And is there just one set of vital goals? A vital goal of a person, I suggest, is a state of affairs that is necessary for this person's minimal long-term happiness. As a consequence of this interpretation many of the things that human beings hope to realize or maintain belong to their vital goals. More precisely, most states that have a high priority along a person's scale of preferences belong to his or her vital goals. Examples of such vital goals can be: taking an exam, getting married and having children, as well as simply maintaining elements in the *status quo* such as retaining one's job and remaining in touch with one's nearest and dearest.

However, certain things that people happen to want do not belong to their vital goals. First we have trivial wants. People may casually want something, but if they don't get it, it does not matter much. Second, people may sometimes have counterproductive wants. They may want to get drunk, but getting drunk is not a vital goal. Instead of contributing to long-term happiness, being drunk contributes in the long run to suffering and thereby unhappiness. Third, we may have irrational wants, i.e. wants that are in conflict with other, more important wants. As soon as the agent realizes this conflict, he or she normally realizes that the only candidates for vital goals are the more important wants.

On the other hand, some things that we do not want may be contained in our set of vital goals. The completely apathetic or lazy person who does not have any conscious goals whatsoever will soon realize that this creates suffering for him or her. This will be particularly salient if the person does not even seek food or shelter. It must certainly belong to this person's long-term minimal happiness to have these basic matters organized. Therefore, such basic goals are among every person's vital goals.

A crucial observation to be made here, then, is that a vital goal of A need not be wanted by A at a particular moment. The notion of a vital goal is thus a technical notion partly distinct from the ordinary-language notion of a goal. (For a further discussion, see Nordenfelt, 2001.)

I will now turn to the idea of health as a second-order ability. To be healthy, I propose, is to have the second-order ability to realize one's vital goals. Consider the following situation. A refugee from, say, an African country, has just moved to Sweden. In his native country he had his own business, which he managed well enough to sustain himself and his family. When he enters Sweden he is no longer able to lead such a life. He does not know Swedish culture and, in particular, the Swedish language, so he cannot initially make any arrangements for establishing a business in Sweden. Whereas in his home country he lived relatively well, in Sweden he is disabled. But would we say that this man is healthy in his native country, and becomes ill upon entering Sweden? No, it seems more plausible to say that as long as he has the second-order ability to run a business in Sweden, then he remains healthy. This means that as long as the immigrant has the ability to learn the Swedish language and the ability to learn how to go about in our society, then he is a completely healthy person. In general, then, such disability as is solely due to lack of training is not an indication of illness. There is reason to speak of illness only if the acts of training have in turn been prevented by internal factors, in which case there is a second-order disability.

But what about the typical case of illness that is due to an organic disease? Consider the following. A woman has a first-order ability to perform her professional activities. Then she becomes ill, and as a result loses her first-order ability. But would it be true to say that she no longer has the second-order ability to do her work?

It is easy to be misled here and identify two pairs of concepts that should be held distinct: one pair is first- and second-order ability, the other is power to execute a basic competence and having a basic competence. We normally ascribe a basic competence to someone when he or she knows how to do something. According to our previous definition this need not at all be true of second-order ability. The immigrant to Sweden has not previously learnt anything about Sweden and does not have the basic competence requisite for making his living in Sweden. He may, however, have the second-order ability with regard to the same action.

It is crucial to recognize that a person who has a basic competence *vis-à-vis* a certain action F need not even have a second-order ability with regard to F. Consider the case of a professional footballer who has broken both his legs. Obviously, during the ensuing period this person does not have the first-order ability to play football. Still, we would say that the person has throughout the period of illness a basic competence to play football. He or she knows how to play football. But does this person, while lying in bed, have the second-order ability to play football? No, for having the second-order ability to play football means having the first-order ability to follow a training programme that leads to a first-order ability to play football. But the person who is confined to bed is clearly not in a position to follow such a programme; and so we may say of the footballer that he or she is ill. The same reasoning may be applied to all paradigm cases of illness due to disease or impairment. During an acute phase of illness, however short it may be, the subject has lost both the first- and the second-order ability to perform the actions with respect to which he or she is disabled.

To this analysis of ability must be added a few remarks about the *circumstances* under which a person can be said to have an ability. It is evident that health cannot be the ability to reach vital goals in all kinds of circumstances. If that were the case then nobody would be completely healthy. There is always some conceivable circumstance in which one cannot reach one's vital goals. The outbreak of a natural catastrophe is one example. Another such circumstance is that a person is physically or legally prevented by other people from performing the actions necessary for the achievement of his or her vital goals. Nor could health be constituted by ability to realize one's vital goals given merely one kind of circumstance. If that were the case then almost everybody would come out as completely healthy. Consider the case where an individual is almost completely dependent on the help of somebody else in his or her endeavour to achieve a goal. We can imagine a paraplegic person who is supported in his or her attempts to reach various destinations; a personal assistant may help in various ways and transport the person. If this were a case where it is true to say that the paraplegic person has the ability to travel to all necessary places, then we should ascribe health to him or her. This is clearly counter-intuitive. Such a situation of extreme support is not one in which we assess a person's degree of health.

So how should these circumstances be defined? A first plausible idea is that the circumstances that we normally have in mind in a health assessment are such as are in some way *standard* in our culture. A person who cannot walk on an ordinary pavement is certainly disabled with regard to a standard situation. Likewise, to take an animal example, a dog that cannot run on an ordinary well-kept lawn is disabled. In both cases, unless there are other impediments, we can draw medical conclusions. The person and the dog are unhealthy.

The way we devise such a standard (which is normally done implicitly) is not via statistics. What is a statistically normal situation (in a geographical region at a particular time) may turn out to be an *unreasonable* situation. In certain countries the political and cultural situation may be such that it would be unreasonable to judge the health of its inhabitants given this situation. It may, for instance, be impossible to work as a teacher in Chechenia for a long time. But it would be unreasonable to say that the trained unemployed teacher in Chechenia is unhealthy for this reason. The circumstances in Chechenia are in this case unreasonable.

On the basis of this reasoning (which can in many details be transposed to the animal world) I have in some texts (Nordenfelt, 2000, 2001) used the term 'reasonable' or 'accepted' circumstances, instead of the term 'standard' that has been adopted here.

Consequences with Regard to Maladies

There is a special conclusion to be drawn from this approach to health and illness, in contradistinction to the biostatistical approach. Although I have above underlined the differences between the two approaches, one must also observe that their respective conclusions are not totally at odds. Many of the diseases picked out by Boorse's biostatistical theory would also be picked out by a holistic theory. A cancer is a disease for Boorse, as well as for myself. But the reasons differ: for Boorse a cancer is a disease because it makes a statistically subnormal contribution to the subject's survival, whilst for myself a cancer is a disease because it tends to create disability (and normally indeed suffering) in its bearer.

For the moment I will disregard the differences between the reverse theories (sometimes also called holistic theories). I think *all holists* would agree that illness has its conceptual root in the notion of a *problem* with one's body and mind, and that, conversely, health is a state where there are no such problems and where one's life has a certain high quality.

But what happens to the notions of disease, injury and defect in this theory? I have in my historical sketch already indicated this. Diseases, injuries and defects are causes or potential causes of the reduced health of the individual bearers. But there is a relevant question here. How can we reconcile the holistic idea of health with the science of diseases dealing with general types such as 'common cold', 'cancer' and 'tuberculosis', when there is some difference between the vital goals of individuals? Can a type of condition sometimes be a disease and sometimes not? How can a common cold be a disease if it does not always reduce the health of its bearer?

My proposal as an answer to this question is that a condition is a disease-type if, and only if, it is an internal type of process of people, most instances of which (but not necessarily all) actually compromise the health of these people. This can be interpreted in two plausible ways, the first of which is the following. Most but not all instances of the common cold actually compromise the health of an individual P. P sometimes has a cold which is hardly recognizable. In such cases we would still say that P has a disease, but that his health is not affected.

In the other interpretation, which is crucial for the science of diseases, we consider the whole population, and say that a condition is a disease-type if, when occurring in many people, it would reduce the health of most of these people. Thus, whatever the vital goals of the members of the population the condition would reduce the health of most of its bearers. The notion of disease would then incorporate a statistical element. This is very different, however, from the statistical element included in the biostatistical concept of disease. In my case the statistical analysis concerns people's ability to reach vital goals; in Boorse's case it concerns the function of bodily organs in relation to the survival of the individual and the species.

From this follows that we can have a science of diseases and still allow for individual variations in the health of people that is dependent on their vital goals. Both P and Q can have colds that in traditional medical terms have identical descriptions. Only P, however, has reduced health, since the cold happens to affect mechanisms relevant to the realization of a vital goal of his, but the same cold does not affect Q's ability to realize her vital goals.

But can we rely on frequency in order to determine the list of diseases? Do people have vital goals similar enough to allow the science of diseases? I think we can be reassured by making the following reflection. A great many maladies, probably the majority, cause pain, fatigue and general unease. Such sensations and moods have a tendency to affect *all* kinds of activity. Thus, the differences in people's vital goals will play a small role. Both if you are an athlete and an author you will be severely affected by pain and fatigue. You will be prevented from performing in the way you had expected and hoped for. Thus the conditions lying behind the pain and fatigue would be classified as diseases.

Introducing a Holistic View of Illness and Health into Animal Science

Is it then possible to apply an analogous reasoning in the animal context? Can we talk about illness, disease and health in a holistic sense with regard to animals?

A first crucial difference between humans and animals is the following: there is among animals no subject that can consciously present a problem using an illness language like the human one. Moreover, there is no animal subject approaching a doctor and explaining what the problem is. Can the holistic concept of illness, then, be the core concept in the animal context?

Let me try to transpose my argument to the animal world. It is true that most animals (as far as we understand) do not embrace a fully fledged language. On the other hand, most other elements present in my human story can be present in a parallel animal story. Most animals can have problems. Many of them, I argue, supported by contemporary animal science, can suffer and can express their suffering. In their wordless way they can ask for help. If the animals in question are in close contact with humans, which is the case with pets and livestock, the humans can interpret the call for help and can try to respond to it. If the humans suspect serious illness, they call for further support and approach a veterinary surgeon, who will act very much like the human doctor in searching for an underlying disease responsible for the illness of the animal.

I have argued with regard to the human case that it is unreasonable to define disease or pathology simply in terms of malfunction in relation to survival and reproduction. I think the case is similar on the animal side. Animals, like humans, can be ill without there being any threat to survival and reproduction. They can be ill in the sense that they feel malaise and have a reduced capacity, as humans can. We can easily observe when our pet dog is not feeling well, when it drags its tail or even whines from pain. A horse can have a disease that reduces its capacity to run in a race. This is a reduction in relation to a goal, but this goal is not identical with the survival of the horse.

But can we make this move as easily as this? What about the existence of animal feelings? There are, as we have seen, a few animal scientists who doubt the possibility of animal feelings altogether (Bermond, 1997). On the other hand, the overwhelming majority of veterinarians and animal scientists base their whole work on the presupposition that many animals can feel pain, fear, discomfort, nausea and fatigue in much the same way as humans.

But can animals feel anything but simple sensations? Is there any point in introducing the more complex vocabulary of feelings? My answer to this is affirmative and I will argue for this position in some detail in the next chapter about happiness and welfare.

One may seriously doubt, however, that *all* animals have cognitions, sensations and emotions. What about worms, cuttlefish and indeed amoebae? Although I think that they must all have some minimal perception (i.e. some mechanism to get information from the outside; see my argument about intentionality below in this chapter), it is probably true that the most primitive ones lack most other mental properties. For one thing they have only a rudimentary neurological system or, in the case of amoebae, none at all. But such an observation is a problem only to those holistic theorists who base their idea of illness *totally* on suffering. Higher animals may suffer, but not all animals do. A theory of health and illness regarding these lower-level animals cannot depend, then, on the notion of suffering.

Animals and Intentionality

What, then, about the ability theories of illness and health? Can they be of further help? Consider, for instance, my own suggestion that health is constituted by the person's ability to realize vital goals. One might suspect that we cannot just transpose my ability theory of health to animals. The main reason is that my abilities primarily regard intentional actions. Crucial such actions are: actions related to daily living, actions related to the subject's occupation, actions related to communication and close human relations, as well as some leisure activities. Most of these are from certain points of view peculiar to humans; a few of them can, for fundamental reasons, only be performed by humans. This holds in particular for conventional actions presupposing a social position, such as being an administrator or a judge.

However, the ability theory of health does not concretely specify the vital goals. Both common sense and animal science clearly show that animals have goals. Consider the following story about our pet dog. Assume that I am engaged in playing with the dog. I raise my hand; the dog observes my movement; the dog sees that I throw the ball; the dog wants to have the ball in its mouth; this is the reason why the dog immediately runs after the ball and seizes it. We see that there is observation and there is a want on the part of the dog, and the want functions as a reason for the dog, and as a result there is a doing triggered by the want and there is a goal achieved. This is a description completely on a par with a description of a boy who is playing with a ball, and it indicates action for a reason. Nobody could seriously deny the validity of this language. We also see immediately how this kind of description can be extended to other mammals, including cattle and sheep, and also to birds. It would not be difficult to describe the behaviour of a hawk hunting a sparrow in the same language of goals.

Many animal theorists claim that the model of animals as stimulus–response automata has largely been replaced by that of animals as goal-seeking systems (Toates, 1987). Researchers such as Dawkins (1990) and Duncan (1996) have devised sophisticated experiments putting animals in situations of choice in order to determine the goals of the animals. They have demonstrated that many animals can make informed choices based on earlier experiences. We can, they say, ascribe a want faculty to most animals. A predator wants to get its prey; the predated animal wants to escape. All animals (that reproduce in a sexual way) want to mate and have offspring, etc. But the animal can also relate to its own sensations, emotions and moods. The cat that has an itch in a leg wants to get rid of it, and therefore scratches the leg. The animal in pain focuses its attention on the pain and wants to get rid of it.

Moreover, it has also been shown that low-level animals with a limited neurological make-up can learn from experience. There are recent remarkable findings concerning fish and invertebrates with regard to the faculties of cognition and learning (Bennett *et al.*, 1996; Karavanich and Atema, 1998; Bshary *et al.*, 2002).

But the sceptic may ask: can one really ascribe agency to animals other than humans? It may be reasonable to use a language of goals with regard to animals. Animals are, one may claim, goal-directed systems. But are animals agents in the sense that they have intentions?

Here I think that a few distinctions are necessary. Let me first distinguish between conscious and unconscious intentions. It is possible that only humans may have conscious intentions in the sense that they are aware of and can identify their own intentions. However, most humans are unaware of their intentions while having them. The man who drives a car from his home to his work is rarely conscious of his intention to drive to work, let alone conscious of the various sub-goals that his driving contains, such as getting to place A before getting to place B. However, there is no question of denying the intentionality of the man's action of driving. If he were questioned about where he was heading he would reply that he was on his way to work. Thus the fact that animals are unaware of their goals is no argument for saying that they cannot have any intentions.

My second distinction is related to the first and concerns the difference between intentionality and capacity for deliberation. The latter is an advanced property. Probably only humans and the higher mammals can deliberate in the sense that they can weigh alternatives and as a result make a decision (which is normally, but not necessarily, a conscious one). Again, the capacity of deliberation is not a necessary condition of intentionality. Many humans rarely deliberate before taking decisions but they are clearly agents all the time.

I have earlier (Nordenfelt, 2000) proposed a dispositional analysis of intentions. Part of the meaning of the locution 'A intends to bring about P' is that A is so disposed that he or she will bring about P, if A is capable of doing so and not prevented from doing so. Another part of the concept relates to the subject's *beliefs*. In order to bring about P, A must also know or believe that certain specific actions are necessary to reach P. Thus an intention is always informed by the agent's perceptions and beliefs. To say that A intends to bring about P is tantamount to saying that A is prepared to perform all the actions that A considers necessary for the realization of P.

In a similar spirit, when we say that an animal has a goal we mean that the animal is disposed to act or move in the direction of this goal. A cat is observing a rat and forms the goal of catching it. As a consequence it starts chasing the rat. The cat is prepared to perform all the actions necessary to catch and eat the rat. The first act is the chasing act.

We also mean, and this holds for all animals down to the most primitive ones, that this goal-directedness is sensitive to new information. When the hungry cat aims at its prey it cannot always just pounce or run directly towards it. The cat may first have to struggle with some competitors that are aiming at the same rat and it may have to avoid other obstacles that stand in its way. All the time the cat uses its senses in order to find the necessary and the most expedient ways to realize its goal.

With regard to higher animals we clearly use the language of cognition. We say that a horse or a cat *knows* or *believes* that something is the case. It is not so natural to ascribe belief or knowledge to animals at the level of worms or crayfish. But these animals, like most other animals, clearly have some sensory input, which in many instances can be stored, and which helps direct their doings and movements in the event of any danger or hindrance that may prevent the goal fulfilment.

In sum, my argument says that most animals, arguably all animals, fulfil the requirement of intentionality and thereby agency. Given my dispositional analysis of intentions, which includes a requirement of sensitivity to information but which does not require consciousness, it seems as if most animals qualify as intentional agents.

But a formidable questions remains. What about *vital goals* in my technical sense? Neither a dog nor indeed a worm embraces the concept of a vital goal. No, but neither does a baby or a senile person, nor for that matter most other people. The notion of a vital goal is a theoretical notion; it is not tied to any hierarchy of preferences in a psychological sense (although I contend that most vital goals are also strongly preferred in a psychological sense). A goal is vital to an animal, according to my theory as it stands, if, and only if, its realization is a necessary condition for the animal's long-term happiness. And there is nothing unrealistic in using these concepts with reference to a large part of the animal world.

But what is included among the vital goals of animals? Again, there need not be much difference from the human situation. There are certainly activities of daily living associated with most animals. Animals take care of their hygiene; most of the wild ones build their own lairs and nests and are quite efficient in feeding themselves. If they were to be prevented from performing these tasks they would suffer. Most animals also have the vital goal of mating and having offspring; some of them are also dependent on living in close communities. These goals are similar to the human ones with regard to sexuality, having children and having close human relationships. The only salient difference between humans and animals is the specific human idea of doing a job for a salary. But this is in reality only a special mode of subsisting, of securing one's survival. The animals still do it in the fashion that we used to employ some thousands of years ago.

The rest of the conception of holistic health in my version is easy to transpose. The idea of circumstances is as necessary for animal action as for human action. All animals act in some environment. But they cannot act in all environments. When we ascribe health to an animal, we presuppose that this animal acts in some reasonable environment (excluding directly preventing ones such as an earthquake or a sandstorm). And when we presuppose such an environment we presuppose what I call a standard environment.

The Problem with Non-vertebrates

But again of course a serious problem crops up. I may have argued convincingly for the case that vertebrates can have vital goals in my technical sense. But what about non-vertebrates? We may perhaps say about all of them that they have goals and that they pursue goal-directed, intentional, doings. But how can they have *vital* goals in my sense, since such goals are, per definition, related to long-term happiness? How do we characterize the happiness of a worm or an amoeba?

My answer to this is that we need for the present purpose a more general notion than that of happiness. The term I propose is that of *welfare*. (I am thereby extending my use of the term 'welfare' beyond what I have done in earlier works, in particular Nordenfelt, 1987/1995.) And in order to substantiate the idea that non-vertebrates can have welfare, let me turn to the philosophy of nature presented by Paul Taylor (1986).

Taylor presents what he calls the biocentric outlook in order to develop an environmental ethics that is not dependent on human interests. This outlook contains the following elements:

1. From the perspective of the biocentric outlook one sees one's membership in the Earth's community of Life as providing a common bond with all the different species of animals and plants that have evolved over the ages.
2. The biocentric outlook also includes a certain way of perceiving and understanding each individual organism. Each is seen to be a teleological centre of life, pursuing its own good in its own unique way. Consciousness may not be present at all, and even when it is present the organism need not be thought of as intentionally taking steps to achieve goals it sets for itself. Rather, a living being is conceived as a unified system of organized activity, the constant tendency of which is to preserve its existence by protecting and promoting its well-being (a summary of Taylor, 1986, pp. 156–158).

In my discussion about feelings in the chapter about animal minds I admitted that we have little evidence for talking about feelings in the case of animals with rudimentary or non-existent neurological make-ups. Here I also think that the holistic characterization of health must use a more generic concept. When the goal of a worm is frustrated, then the worm, as far as we know, does not feel pain. However, something negative has happened to this worm. The welfare of the worm or the quality of the worm's life has been reduced. We can infer this from certain behaviours or non-behaviours on the part of the worm. It may make stereotyped and unsuccessful movements, with regard to attaining a particular a goal, for instance crossing a path in order to reach an attractive heap of leaves. Or it may not move at all. We may also compare this worm with another worm crawling just beside it. The latter worm is lively, it moves around quickly, it is bigger and it seems to be thriving. The first worm, we say, must have some problem. It is ill.

I can, indeed, also tentatively approach the universe of plants and the matter of how we might extend a holistic approach to the understanding of the health and illness of plants. The notion of happiness is clearly not applicable to plants. On the other hand we may be able to talk about a reduced quality of the plant. We can compare the life of a particular plant over time. One year it flourishes and grows a lot. Another year it shrinks and even withers and does not produce any flowers. A third year the plant is revitalized and is flourishing. When it flourishes the quality is high and the plant is well; when it withers the quality is low and the plant is ill.

My suggestion for the systematic characterization of health and illness in the case of lower animals and plants is then the following. *An animal A or a plant P is healthy if, and only if, A or P has the ability to realize all its vital goals given standard (or reasonable) circumstances. A vital goal of A's or P's is a necessary condition for the long-term welfare of A or P.* In the case of the higher animals the criterion of welfare is the happiness of the animal. In the case of non-vertebrate animals and plants we must find other criteria, such as the vitality of the animal or plant. Observe also that in the case of plants we cannot talk about actions or doings. The ability of plants refers to growth and development. (For an interesting comparison between animals and plants with relevance to this issue, see Lopez *et al.*, 1994.)

Two Problems with a General Welfare Approach to Health

Here I must try to tackle two crucial problems with my reasoning. The first problem is an argument which says the following. When we are talking about non-vertebrate animals and the plants the only goals that we can find are the biological goals of survival of the individual or survival of the species. There are no other goals that could be termed vital among these living beings. Thus the biostatistical theory of health must come out as the only viable theory in this domain.

I first wish to dispute the general validity of this statement. It seems highly implausible that *all* strivings and all movements of non-vertebrate animals have the sole purposes of surviving and reproducing. And if we admit that there may be at least some striving of such an animal that has another goal and that this striving can be frustrated by some inner process in the animal, then we have a case of illness that need not be explicable in biostatistical terms. Similarly, it seems highly implausible to say that the only highly valued quality of a plant is its development for the purpose of surviving and reproducing itself. And if there are other highly valued qualities and they can be reduced by inner processes of the plant, then there is a case for talking about plant diseases in other terms than biostatistical ones.

On the other hand, to cut the debate short, I can, for the sake of argument, admit that to the most primitive living beings, such as the single-cellular ones, one cannot reasonably ascribe any further goals than the ones of being able to survive and to reproduce. Thus the diseases of such creatures would be well covered by the Boorsian concept. Hence in such cases there is a collapse between vital goals and certain biological goals. From this, however, does not follow that the holistic theory fails to work in these cases. It is only that holism in these cases contains only certain biological goals. The welfare of a single-cellular organism may be constituted only of its ability to survive and reproduce.

I think this point has to be emphasized. Survival and reproduction are not vital goals simply because of the biological fact that the organisms have a tendency to strive toward these states of affairs. Survival and reproduction are vital goals, as I see it, because they are inherent *goods* of the organisms in question. Thus although there is a factual overlap between the biostatistical and the holistic viewpoint on this matter, there is a fundamentally different interpretation of the facts.

The second serious problem stems from a fundamental question. How can we determine the existence of a problem in a low-level animal or a plant? What is such reduced welfare of a low-level animal or a plant as is not identical with a lowered capacity for the survival of the individual or the species? Is there not a great risk of gratuitous ascriptions of goals and qualities to these creatures? Yes, there is such a risk. There is the particular risk of implanting *human purposes* in such animals and plants and indeed also in higher animals. Such health language already exists among veterinarians and agronomists. An animal or a plant may be deemed unhealthy because it does not fulfil the human expectations with regard to production, for instance of meat, milk or crop.

Indeed, when this health language is used, as for instance when we ascribe illness to a cow because it does not produce enough milk for us humans to consume, then we use a health conception that is holistic. On the other hand this is not what I, and probably not what most of my holistic colleagues, *primarily* have in mind. Welfare, according to us, must be understood from the subject's point of view. A vital goal, in my theory, has to do with the subject's welfare. This is quite clearly indicated in the case of the higher animals, for which the crucial concept in the theory is the subject's long-term happiness.

In my preliminary discussion of animal and plant health above I was using the term 'vitality' in an attempt to cover the welfare of animals and plants. The Finnish philosopher Von Wright (1963, 1995) in a treatment of the concept of need in fact borrows the term 'flourishing' from the plant world to be applied to all living beings. A need of a living being, he says, is related to its ability to flourish, and not simply to its ability to survive. And, I would like to add, when such a need is frustrated because of processes internal to the animal or plant, then this creature is ill.

But how do we determine that vitality or flourishing is present from a subjective perspective when we have no real access to a subject's point of view? I do not deny the difficulty and profundity of this question and I have no illusion of coming up with an ultimately satisfactory answer. However, for the purposes of this book there is an efficient short cut. I think my argument shows the following: to the extent that we are inclined to talk about the welfare of animals and plants which is over and above their sheer ability to survive and reproduce, we have a case for holistic notions of health and illness. Many people, including myself, are willing to ascribe vitality or flourishing to animals and plants. We sometimes also observe how this vitality or flourishing can decrease. On these grounds we can ascribe illness to the animal or plant. Our concept of illness is then a holistic one.

Concluding Remarks

Let me conclude this analysis. I have argued that, perhaps contrary to common belief, the so-called holistic theories of health are not confined to the human arena. We need not say that there is something special about human health that calls for a holistic theory of health, whereas the biostatistical theory or some equivalent theory must do the work in the case of animal health. I argue that health and illness could, or indeed should, be viewed in a holistic way with regard to all the living world. The concept of illness, I maintain, is related to the notion of a *problem* caused by internal processes. And all living beings may have such problems. These problems may be a threat to survival and reproduction as the biostatistical theorists say. However, they need not constitute a threat of this kind. The problems can be related to other goals. Hence there is a case for a holistic theory.

A further crucial consequence of this argument is that I am proposing a notion of health that is connected to the notion of welfare. The two key notions of this book are thus tied together. I have already suggested how welfare could be construed in my system. However, a more detailed investigation is warranted.

22 Towards a Happiness Theory of Welfare in the Animal Context

A General Background

Humans and animals all live in some kind of environment. This environment has many parts. First, there is a physical environment, a landscape with natural resources and climate. Secondly, there is a cultural environment, a society with laws and regulations, a political system, customs and other expressions of culture. The cultural environment is particularly salient for humans but it also has a great impact on animals, not least with regard to regulations for human interference with animals. Thirdly, there is a close psychosocial environment consisting of relatives, friends and co-workers in the human case, and of mates, offspring and human carers in the animal context.

This environment influences our lives in many ways, when seen from a logical point of view. First, there is the direct *causal* influence. There are fundamental positive causal influences from our physical environment; we get our nourishment from nature and we get the necessary warmth from the sun. The climate has a direct physical impact on our bodies, both positively and negatively. The artefacts of human civilization, constituting the cultural environment, are often important positive factors. Houses, for instance, provide us with shelter and protection. Other products of civilization, however, are detrimental. Chemicals in the water and in the air can and sometimes do have disastrous consequences for humans and animals.

But the environment is also the platform for our actions. It gives us the opportunity to indulge in various activities. These opportunities vary enormously in different parts of the world. Greenland, for instance, provides the opportunity to go out hunting and fishing but it gives very little opportunity for agriculture. Southern Europe makes it possible to cultivate grapes and citrus fruits, but that part of the world is not suitable for raising reindeer. Scandinavian society gives the opportunity to express one's opinions and establish a variety of political parties. This is currently not the case in Syria and North Korea. We can easily multiply such examples from many sectors of human life.

Mutatis mutandis we can specify the variety of biotopes for animals. The Arctic countries make a certain kind of living possible. They are suitable for species that can tolerate a harsh climate, are dedicated to sea life and take most of their nourishment from the water. The African rainforests are suitable for all the animals that require warmth and an abundance of vegetation and the presence of other animal species.

While being a platform for action the environment also sets the limits for *our goals in life*. We cannot seriously want to achieve anything in every environment. We cannot create a fleet of ships in Hungary and we cannot build a capital on the slopes of Mount Everest. This certainly also holds on the individual level. I understand that I cannot create a vineyard in my Swedish garden but must confine myself to growing apple trees. Similarly, the birds in my garden choose to build their nests using the only material that they see is available.

The environment provides opportunities and restrictions in terms of both being a platform for action and being a basis for setting our goals. The two ways are related but can be distinguished in the following way. In the former case the environment is the physical and cultural space for behaviour; in the latter case the environment is *perceived* by the agent and on the basis of this perception humans and animals choose to set their goals.

Thus the whole many-faceted environment influences us physically and mentally, both in the short and in the long run. It affects in particular our welfare. The environment is an extremely important, although not the only, foundation for our welfare or illfare. (In my earlier presentations referring only to the human case – Nordenfelt, 1993 – a slightly different nomenclature was used. For a commentary on this, see the Introduction in this book.)

In the light of this analysis I propose a distinction between the external conditions of welfare and the welfare itself. A state of welfare is created, affected or annihilated by different combinations of external conditions. To the effect that such a combination of external conditions contributes positively to a person's or animal's welfare we call it a state of *external welfare*. To the effect that it contributes negatively to the welfare of an individual, or even creates illfare, I call it a state of *external illfare*. In what follows I shall mostly refer to the positive case. The simple terms 'welfare' and 'illfare' will in the following denote the internal state of welfare of a human or an animal.

This characterization must now be supplemented. There is an important domain between external and internal welfare. There are not only external conditions of inner welfare, but also a series of internal conditions. It is not only the external world that affects us. We are also, to a great extent, determined by our own physical and mental *constitution*: how we are constituted of physical and mental elements, our physical and mental strength, our health and our character, as well as our inclinations and interests. Our inner properties are, certainly, continuously affected by the external environment, but there is an important constitutional basis that the environment cannot affect in any other sense than that it can annihilate it.

I will use the term 'internal conditions' to refer to the internal state that together with external factors determines our state of welfare. A person's or animal's health is one of these conditions. Thus, I look upon health as an internal condition for welfare and not as a part of welfare. This is evident since health is defined as a person's or animal's ability to achieve welfare given certain circumstances.

Still, my preliminary analysis is not sufficient. There is an area which is a product of the outer and inner conditions of an individual, and which to a great degree affects the individual's welfare. This is *activity* itself. If we reflect for a moment we can see that the most typical way in which outer and inner conditions exert influence is through our own activity. In order to be affected by some state of affairs we must often act upon it or react to it. In order for most resources to lead to welfare they must be *utilized*. In order to get crops we must go out and cultivate the earth. In order for a tool to be appreciated it must be used. We must use our weapons and fishing rods in order to get meat and fish. Similarly, the predators in the animal world must take advantage of an opportunity and use their strength and skills in catching their prey.

Sometimes perception is sufficient for the external world's affecting us. Our welfare can be affected simply by our realizing that something is the case. One can become happy about the fact that one has inherited the sum of £100,000. One can feel grief at the famine in Africa. An owl can become excited at the sight of a rat. Perhaps we can say that this is a case of influence without the involvement of activity. But this is a qualified truth. In order for influence to occur in this kind of case it is required that the subject perceives and even *reflects* upon the facts. The person who has inherited must be aware of the positive consequences of money in order to react positively. Similarly, in order to get excited the owl must understand that the rat can function as a tasty breakfast.

Let me sum up in a preliminary way. Both the external and internal environment of a person or an animal can, together with his, her or its own activities, affect the person's or animal's welfare. In addition, there is a series of both logical and empirical connections between these main categories. The environment can influence the constitution, the environment together with the person's or animal's constitution can influence the chosen activities, and so on.

Consider now the relation between the concepts of external and internal welfare. By the external welfare of a particular individual I mean the compound of things in the external environment that together with the individual's internal conditions and activities influence his or her welfare. A presupposition for a condition's belonging to the external welfare of the individual is thus that it contributes to the welfare of this individual. The set of states constituting the external welfare of an individual P is thus not established in some independent way without any connection to P's (internal) welfare. The relation between the concepts of external and internal welfare is thereby *conceptual* and not empirical.

An immediate consequence of this is that the conditions for A's welfare need not be identical (not even in the sense of type-identity) with the conditions for the welfare of A1. The conditions for A's having a good time can be distinct from the conditions for A1's having a good time. Those things that make life good to a banker in New York are at least partly different from the things that make life good to a housewife in Helsinki. The good life conditions of a sea lion are different from those of a cow in Scotland or a tiger in India.

The sets of conditions for welfare are thus partly different. But it is also clear that they have a common core. The conditions of pure survival are necessary elements among all conditions of welfare. Thus such conditions (including some degree of health) belong to the welfare conditions of all people. Most probably some further conditions belong to the common kernel of all humans and animals. It is quite probable, for instance, that a certain minimum of social relations have a place here. (The answers to these questions certainly also depend on how abstractly we formulate the conditions.)

Towards an Analysis of Welfare and Happiness

My first point of departure is that the concept of welfare covers the whole area of positive human and animal experiences, from *sensations* to *emotions* and *moods* (Nordenfelt, 1993). It is customary in modern philosophical psychology to differentiate between these categories (Kenny, 1963; Goldie, 2000). Sensations are such feelings as have a clear bodily location and as lack a direction. Typical examples of sensations are pains and itches. When one feels pain one feels it at a particular place on or in the body, for example one's knee. When there is an itch it occurs at a particular place, for instance behind one's ear. Emotions and moods do not have this bodily location. When one is in love or is frustrated (cases of emotion) one does not have a feeling that is situated in any particular part of the body. Nor does anguish (a case of mood) have a specific bodily site.

The latter statements should be qualified in order not to be misunderstood. What is intended is that one cannot identify the emotions of love and frustration, nor the mood of anguish, at a particular place in the body. Emotions and moods are more complex states of affairs than locatable sensations. This is not to deny that there may be sensations typically associated with emotions and moods. A sensation of elevation in one's breast is perhaps associated with love. A sensation of physical fatigue may be associated with frustration. And a sensation of pain in the stomach may be associated with anguish. These sensations will often vary with individuals and, what is crucial, they are not identical with the complex emotions of love and frustration and the mood of anguish.

It is also crucial to differentiate between emotions and moods. The feature distinguishing between these two is that emotions are directed towards objects, normally outside the individuals themselves, whilst moods have no such direction and lack objects. The class of emotions is comprehensive, including love, hatred, joy, sorrow, hope and despair. When one is in love, one is in love with *somebody*. Similarly, one is sorry about *something* and one hopes for *something*. (For the most detailed analysis of such features, see Kenny, 1963.) As examples of moods we find calmness, (certain instances of) depression and anguish. These lack objects and they are not located in any specific part of the body. It is interesting to see that the emotion/mood distinction proposed in philosophical psychology is similar to the one suggested by the animal scientist John Webster (1994, p. 28): 'Mood defines a state of mind, such as anxiety, that cannot necessarily be linked to a specific stimulus, nor interpreted as a reason for a specific action.'

In order to avoid misunderstanding again, I will underline that these theses about the location of emotions and moods are not meant as theses concerning the neurophysiology or biology of feelings. They concern the phenomenology of feelings; they concern how we as ordinary human beings look upon our feelings and how as a result of this we use the words of feelings in our language. To say, for instance, that we do not locate a feeling of love in a particular part of the body is therefore completely compatible with a scientific hypothesis that there are certain locatable neurophysiological and endocrine processes involved in a particular feeling of love. (See my earlier presentation of contemporary neuroscience in Chapter 12.) It may, however, very well be – as is suggested by my observations above – that there is no specific set of neurophysiological or endocrine processes that is involved in all instances of love. Given this hypothesis one cannot identify love solely through inspection of the biology of man; the identification of love can only be made in a context of human interaction.

Can the different emotions be directed to any kinds of objects? An analysis of the emotion concepts shows that there are quite clear limitations. One can in fact give a salient demarcation of the kind of thing that can function as the object of a specific emotion. For instance, one cannot hate a person unless one believes that this person has performed or is going to perform terrible deeds that are highly threatening to oneself or some of the interests of oneself or a loved one. One cannot despise a person unless one believes that he or she has behaved in a cowardly or otherwise morally reprehensible way. One cannot hope for something unless one believes that what one hopes for is something good and that there is a chance of getting it. The type of object that an emotion can have is called *the formal object* of this emotion (Kenny, 1963).

These statements are not conclusions from empirical observations. They are derived from the logical grammar of the concepts involved. A person who uses the terms 'hate', 'despise' and 'hope' in other ways cannot be understood unless he or she gives completely new definitions of these terms.

On Feelings in the Animal World: a Recapitulation

Are these distinctions among feelings relevant to the animal world? Can animals feel anything but sensations, such as pain and fatigue? Is there any point in introducing the more complex vocabulary of feelings? In the previous section on animal minds and health I have already argued that this is the case. I will rehearse some fundamental points and make some additions here.

Panksepp (1998), from his neuroscientific perspective, first of all emphasizes that most scientific observers are convinced that at least all mammals can
experience states of feeling. It is apparent in their outward behaviours. They
exhibit motivated and avoidance behaviours. Moreover, there is compelling
evidence from psychopharmacological research, where behavioural changes in
animals can predict human subjective responses. Panksepp also describes how
contemporary neuroscience has located or is in the process of locating centres
in the midbrain of mammals that are the main monitors of their affective life.
He identifies at least seven such centres and has coined for them the names:
seeking, fear, panic, rage, lust, care and play.

Panksepp is highly conscious of the pitfalls in using the ordinary language
of human feelings in identifying these neural states. In using such terms as
'panic', 'rage' and 'lust', he is coining a provisional language that has some
connection to our ordinary language. Panksepp does not make the distinctions
between various mind categories, such as that between sensations, emotions
and moods, which I introduced above. On the other hand, the categories he
has identified are clearly more complex than simple sensations of pain and
fatigue. They refer to such mental states as can be aroused by the animal's perceiving a state of affairs, for instance the fear that an animal can feel in the
immediate presence of a predator, or the rage that an animal can feel when a
competitor approaches. These are, in contradistinction to localized 'physical'
pain, feelings that are informed by and prompted by knowledge. This already
gives us sufficient reason for talking about the existence of emotions (in the
sense identified above) in animals, and indeed a great variety of such emotions.

With regard to cognitions, which are not in focus here, a great number of
authors, including Griffin (1984, 2001), Ng (1995), Vorstenbosch (1997) and
Panksepp (1998), have presented rich evidence for their existence. In particular Griffin has argued in great detail, with the help of an abundance of examples, for the existence of complicated thinking among many animals.

Several animal scientists and veterinarians have added to the picture of
neuroscience by referring to ethological or in general behavioural data. One
such animal scientist is Donald Broom (1998), who has characterized a variety
of animal feelings and demonstrated feelings of great complexity, including
both emotions and moods, as explicated above. Broom discusses in detail the
emotions of fear, grief, frustration and guilt in animals. In the case of fear he
uses the following definition: 'a normal emotional response to consciously recognized external sources of danger' (p. 383). He notes that this is an extremely
common emotion in all animals, one that has an obvious role in the animal's
coping process. Concerning grief, Broom notes that there are many reports of
pets, especially dogs and monkeys, that show the same sort of behaviour that
humans show in similar circumstances. 'This behaviour is sometimes described
as mourning and has also been described in horses, pigs and elephants' (p.
385). Broom also quotes specialists who detect behaviour expressing guilt, an
emotion one could have believed to be uniquely human. Dogs may hang their
heads; monkeys may behave in a submissive way and 'grin'. Broom makes
similar observations with regard to frustration and to the moods of depression
and anguish.

A particularly rich source for the study of animal feelings in a welfare context is John Webster's *Animal Welfare: A Cool Eye Towards Eden* (1994). Webster has a broad grasp and starts with the sensations, in particular pain. He notes first, by reference to basic science, that the fundamental anatomy and physiology involved in the processing of information from the noxious stimuli are common to all vertebrates. He observes, second, that sentient animals show learned aversion to painful stimuli. They obviously remember what caused them pain and they avoid the source and site of the cause of that experience again. Moreover, Webster can refer to subtle experiments where devices are used to measure thresholds of pain. The trained animals will flinch as soon as the threshold of pain has been passed. Two other basic sensations that are central in the welfare context are hunger and thirst. They have likewise salient and well-known physiological, neurological and behavioural causes and effects.

A further set of sensations extremely relevant to animal welfare is the set associated with the comfort and security of the animal. One of the five theses proclaimed by FAWC in 1993 is the freedom from discomfort, and this freedom should be achieved by 'providing a suitable environment including shelter and a comfortable resting area'. Comfort is complex and relates to such different aspects of the surroundings as temperature, space, hygiene and security at rest.

With regard to emotions Webster pays much attention to fear, which he finds to exist almost universally among vertebrates. He makes the fruitful distinction between innate fears, learned fears and the fears engendered by the perception of fear in others of the same species. All kinds of fear, he claims, are present among all humans and sentient animals. Fear, although basically a negative feeling, can be adaptive and therefore a good thing. It helps the animal to avoid a threat. The fear becomes, however, purely negative and an element of illfare when the action fails to remove the threat or when the animal becomes uncertain about what action to take.

It is significant that Webster also acknowledges the emotions of frustration and boredom, not often discussed by other authors. He finds that the assumption of such emotions is necessary to explain the behaviour of a horse that has been isolated for a whole day in a stable or a dog that has been isolated for a day in a flat (Webster, 1994, p. 202). He also conjectures that pets may suffer more from boredom than farm or wild animals. There are two main reasons: first, the pets do not work so they have more time to kill; and second, they tend to have higher expectations.

The Different Kinds of Welfare

With the distinctions among feelings as a background we can now return to the different species of welfare. What are they and how should they be classified? As a candidate for a sensation we can find at least one, viz. sensual pleasure. The receptors of the senses, such as smell, taste and touch, can give us sensations of pleasure. In our consciousness we would also locate these sensations in the relevant parts of the body, in the nose, the mouth or some part of the skin.

There are several emotions with a salient positive association. Among these are love, friendship, appreciation and joy. Also the mood category includes several kinds of welfare. To these belong, for instance, calmness, peace of mind and harmony. These feelings have no particular physical place. We feel neither love nor harmony in the legs, the stomach, the heart or the brain.

What, then, about happiness? It is not a sensation. Happiness is clearly not located in a particular part of the body. It is equally clear that happiness can take objects. One can be happy about something. One can be happy about one's progress, one can be happy about one's family life and even about life in general. Happiness is therefore in all these cases an emotion. It could be argued that there is also a mood sense of happiness. This would be the case when we say of a person that he or she is simply happy, without having anything to be happy about. I do not wish to dogmatically rule out this kind of case. However, there may be an object of a person's happiness without him or her being aware of this. Many cases of just being happy could therefore be analysed in terms of unconscious objects.

Happiness has a special position among the species of welfare. (For the human case I have previously argued that happiness is the major candidate for constituting quality of life.) First, happiness is the most general of the concepts of welfare. In a way it can be said to incorporate the others. It can do so through its nature of being an emotion, i.e. by being a feeling directed towards an object. The object of happiness can be of many kinds; for this reasoning it is crucial to note that the object can be *another feeling*, for instance a feeling of the well-being kind. One can be happy about being at ease and one can be happy about an experience of pleasure. Another way of expressing this is to say that happiness is a species of *welfare of the second order*. Happiness is a consequence of one's reflecting upon one's life. One observes and reflects upon some phenomenon in life and as a result one feels happy about it.

On Happiness

Is there then an abstract way of characterizing the formal object of happiness? What can we be happy about? A classical answer to this runs in the following direction. Happiness is conceptually connected to the wants and goals of human beings. (For various formulations of this answer, see McGill, 1967; Tatarkiewicz, 1976; Telfer, 1980.) One is happy about the fact that one's wishes and goals are realized or are becoming realized. If one's life as a whole is characterized by the fact that one's most important goals are fulfilled or are in the process of being fulfilled, then this life is a life of great happiness. It is crucial to emphasize here that the goals talked about need neither be conscious nor be the result of personal achievement or even constitute a change. One of our most important goals is to maintain the *status quo*, to keep our nearest and dearest with us, to keep our jobs and in general maintain our most fundamental conditions in life.

This observation about the relation between happiness and the realization of our conscious or unconscious wants gives us immediately a fruitful suggestion for judging the role of external conditions in the assessment of welfare. I will first look at the matter from an individual point of view. The external states of affairs that have directly to do with my happiness are such as contribute to – or prevent – the realization of my goals. These states of affairs thus constitute my external welfare.

Now, different people partly have different life plans. Some of us have very ambitious and expansive goals; others, the more cautious people, set goals on a lower level and require little from their lives. Thus the external conditions for happiness (i.e. the external welfare) can vary greatly between people. This observation tells us that we can never reliably characterize the happiness of a particular person just by describing certain parts of this person's life situation unless we know about the relation between this situation and his or her wants and goals in life.

With this reasoning as a background I will now give a preliminary characterization of the concept of happiness:

P is in a complete state of happiness if, and only if, P wants the conditions of life to be just as he or she finds them to be.

Or, more formally:

P is completely happy at t if, and only if,

1. P wants at t that $(x1,...,xn)$ shall be the case at t,
2. $(x1,...,xn)$ constitutes the totality of P's wants at t,
3. P finds at t that $(x1,...,xn)$ is the case.

This is then the list of conditions for *being* happy. Being happy is, as I see it, a disposition for *feeling* happy. One can be happy without at the same time necessarily having a feeling of happiness. A happy woman, for instance, can be happy overall about some personal success. This does not entail that she must have a particular experience continuing during her whole period of happiness. There can be a long period when she does not pay any attention at all to the object of happiness. Then, she does not have any feeling of happiness at all. If, however, somebody were to remind her about her success, and she actively began to think about it, then the probability is great that she would also have a feeling of happiness.

Therefore, *feeling* happy is not exactly the same as *being* happy. The two are analytically connected but not identical. Feeling happy entails being happy but the reverse does not hold. To be happy is only to be disposed to feel happy. This is a disposition that is particularly likely to be activated when the subject pays attention to those conditions that constitute the realization of his or her wants.

A question remains. I have characterized happiness in general, i.e. also the pure disposition, as an emotion. Should we not reserve the term 'emotion' for the special case where an experience is present? I have here allowed myself to use the term also for the general case. This follows a tradition in the modern theory of happiness where emotions are analysed solely in terms of beliefs and desires and not in terms of subjective experiences. (See, for instance, Ingmar Pörn, 1986.)

It follows from this characterization that happiness is a dimensional concept. P can be more or less happy with life according to the degree of agreement between the state of the world as P sees it and P's wants. There is a continuum, then, from complete happiness to complete unhappiness, the latter being a state where there is no agreement at all between how P sees life and how he wants it to be.

I will suggest that the opposition between happiness and unhappiness is of the contradictory kind. This means that the continuum can be divided into two mutually exclusive parts, one part of happiness and another part of unhappiness. (In Nordenfelt, 1994, I have tried to characterize the point at which happiness and unhappiness meet.)

To the global notion of happiness with life as a whole corresponds molecular notions of happiness with particular facts. P can, for instance, be happy about the fact that he has passed an exam or that he has received a nice gift from a friend. In a way the global happiness is a function of various molecular 'happinesses'. This sum, however, cannot be derived in a simple arithmetic way. Our general happiness or unhappiness is dependent, not so much on the quantity of matters we are happy about, but on the importance we attach to these matters. To most of us it is more important to become a parent than to spend a nice day in the forest. The father's happiness about his newborn child influences his general happiness much more than his happiness about an excursion in the country.

Some Complications with a Preference Theory of Happiness

There are certainly difficulties surrounding a preference theory of happiness. I have previously dealt with a few of them in Nordenfelt (1993, 2000). I will here focus on two problems. Consider first the problem of *low levels of want equilibrium happiness*.

Can one not have too low aspirations or too small a set of wants? Can the want equilibrium criterion work in the minimal case? Is it reasonable to say of a person who does not want to do anything else than lie on a beach enjoying the sun, that he or she is happy just when this want is satisfied?

To discuss this example in a reasonable way we must specify a few further conditions. Strictly speaking this person *cannot* have only this want. The person must now and again want to eat, drink and sleep, and these wants have to be satisfied. But we must also assume that this person does not frequently get bored and hence in fact quite often wish to do something else and perhaps more useful. Most sunbathers do not, in the end, fulfil the second requirement.

They become terribly bored with life in general and want desperately to do something else. But they may be unable to change their situation for various psychological or circumstantial reasons.

Having said this we must allow for the extreme case of a person (or an animal) who is not bored and who genuinely does not want a richer life. This would presumably be a person or an animal with a low degree of intelligence and emotional sensitivity. But it would be wrong to deny that such a person or animal has happiness.

But consider now another challenge to the theory of *happiness as an equilibrium*. So far I have accepted that a person or animal can be completely happy, viz. in the case where he, she or it has all the wants satisfied. But what kind of completeness is this? Does this preclude a person or animal from becoming happier in the future? And how could this change for the better be theoretically characterized? In a previous text (Nordenfelt, 1993) I introduced a further possible dimension of happiness in order to allow for this possibility. I called this dimension the *dimension of richness*. As an illustration I used the following example.

A boy has lived all his life in simple and unpretentious circumstances in the Highlands of Scotland. He has been entirely content with his lot; he gets along well with his family, he appreciates the wild countryside and he enjoys the sometimes hard struggle for life up there. Thus he has been completely happy in the want equilibrium sense as analysed above. One day he and his family are visited by a tourist who happens to be a famous musician. The tourist is attracted by the place and settles there permanently. Partly to earn his living he starts teaching the young boy to play the violin. He then discovers that the boy has a remarkable talent for music and that he very soon develops a proficiency in playing. This completely changes the boy's life. A whole new world has been opened to him. He enjoys playing and he enjoys listening to music. To put it in my technical terms, he has acquired a number of wants that he did not have before, and he is in the process of satisfying them.

The boy was, as I said above, completely happy with life before the musician arrived. But how should I then express the positive change that has now happened to him? Let me describe the situation formally. A wants at time t the states x, y and z to be the case. x, y and z happen to be realized at t and A is completely happy. At t1 A gets to know about the existence of a new state of affairs e and wants to have that. He still wants x, y and z. Assume now that x, y, z and e are realized at t1. And assume also that there is nothing further that A wants to have. Is A as a result necessarily happier at t1 than at t?

According to my present analysis we cannot in general draw this conclusion. It may be true that A at t1 still wants x, y and z. But – presumably as a result of his strong want for e – x, y and z may have declined considerably in importance for him. Thus it is not clear that A is happier in the later situation than in the earlier one. This difficulty does not entail that we cannot in any situations claim that a person is happier at t1 than at t, although he or she was at t completely happy. We must then as before rely on the notion of preference. The conditions can be formalized in the following way. Assuming that A is completely happy at t, A can still be happier at t1 than at t if, and only if,

1. A prefers at t1 the total situation at t1 to the total situation at t.
2. A would at t prefer the total situation at t1 to the total situation at t, had A at t been adequately informed about what the total situation at t1 would be like.

Thus the notion of complete happiness in the equilibrium sense is, as I now see it, a relative one. The completeness is relative to the set of wants that a person or animal has at a particular moment. It does not take into account the person's or animal's potential wants. From a conceptual point of view there is then *no absolutely complete happiness*. The only limit to happiness, then, is an empirical one. A person or an animal has only a limited set of wants and he, she or it cannot realize more than a limited set of wants.

Happiness and Other Species of Welfare

What is the relation between happiness and other species of welfare? It has already been stated that happiness is a second-order emotion, sometimes taking other species of welfare as its object. One can, for instance, be happy about one's sensations of pleasure. This observation should be somewhat qualified. There is no necessary connection between pleasure and happiness. A person need not always be happy about his or her pleasure.

Consider the case where the pleasure is a sign that something dangerous is going on, the pleasure involved in taking a drug, for example. The individual may be conscious of the fact that after a while the pleasure will be gone and that the future suffering will be great. Hence, although at a particular moment he or she may experience intense pleasure, he or she may at the same time be deeply unhappy.

Conversely, pain and suffering are states of mind that are typically unwanted. Normally, a person in great pain is unhappy about his or her mental state. The pain or suffering may be a sign that something positive is to be expected. A necessary surgical operation may be very painful. However, if the patient strongly believes that he or she will be better after the operation, then the patient may combine the pain with happiness.

Further Applications to the Animal Case

Can happiness and preference be the core notions in the animal case? In a way I have already answered this question in the section about the preference theories of animal welfare. Animal welfare theorists, such as Appleby and Sandøe (2002), have suggested a notion of welfare defined as experienced preference-satisfaction which has great similarities to my own definition of happiness as satisfaction of wants and preferences. Other welfare theorists, such as Dawkins (1990) and Duncan (1996), have devised sophisticated experiments putting animals in situations of choice. They have demonstrated that many animals can make informed choices based on earlier experiences.

In general, it has also been noted that we can ascribe a want faculty to most animals. A predator wants to get its prey; the predated animal wants to escape. All animals (that reproduce in a sexual way) want to mate and have offspring, etc. Many animals can also relate to their own sensations, emotions and moods. The cat that has an itch in a leg wants to get rid of it, and therefore scratches the leg. The animal in pain focuses its attention on the pain and wants to get rid of it. Thus there is also in animals a set of feelings of second order. (In this reasoning I certainly exclude the lowest-level animals; see my discussion above with regard to health.)

In saying this I do not intend to say that the animal has all the qualities associated with human happiness. Most animals cannot reflect on their complete situation in the way many humans may do. Only a few animals can have hopes for the future, let alone predict what their future feelings will be in anything like the way that humans can. Only a few animals can have feelings about their past, for instance feel grief and feel ashamed. On the other hand, we can say about most animals that they are in a state where all their wants and preferences are satisfied. This is indeed all I need in order to apply my 'technical' notion of happiness to these animals.

But can such a notion of happiness be adequate for the purpose of defining animal welfare? Is the happiness of the animal the only goal we wish to achieve? Here I have conflated two questions that are not identical and do not have identical answers. I think that happiness (in my technical sense) can serve as an explication also of much animal welfare (but not all of it; see my analysis of health in Chapter 21). But when we are talking of the *general goals of animal care* there are other things to be taken into account. First of all we must consider *welfare in a sustainable sense*. The momentary happiness of the animal is almost irrelevant if its welfare in the long run is endangered. We may know that the animal has a serious disease which will most probably result in grave future disability and suffering. Then our goal must be to cure the disease if possible. We may also know that there are external factors that may endanger the welfare of the animal. Assume that a decision has been taken to the effect that a group of cattle shall be moved to a biotope that is not at all suitable for them. The external welfare of the animals is being turned into illfare. Then, action ought to be taken to avoid this change.

There may also be goals that humans may have with regard to animals that are not identical with the individual welfare of the animals. The goal of killing animals for food is quite a different goal. It may to some extent be compatible with the goal of animal welfare. Livestock can be kept in a reasonable way, with consideration for their welfare, before being taken to slaughter. But the act of slaughtering is certainly something different. Whatever our ultimate view of the slaughter of animals we must concede that the act of slaughter is the direct opposite of caring for the health and welfare of the animal. Similarly, the goal of feeding animals with the human consumption of their meat in mind is also quite a different goal. To balance such goals with the goal of individual animal welfare is, indeed, one of the fundamental issues in animal welfare ethics. But this is the topic of quite a different research project.

Appendix
On Amartya Sen's Theory of
Functionings and Capabilities

In this appendix I wish to relate my proposal with regard to the notions of health and welfare to the well-known theory of welfare proposed and developed by Amartya Sen in a great number of publications. On the surface there are similarities between Sen's emphasis on so-called capabilities and functionings in his characterization of welfare and my own choice of abilities in characterizing health. What therefore needs to be asked is whether my notion of health is closely similar to Sen's notion of welfare, and what conclusions, if so be the case, can be drawn concerning the reasonableness of the two theories? Moreover, the idea that welfare is constituted by a person's (or an animal's) functionings and capabilities is one that has not been scrutinized so far in this study.

Sen's theory of welfare can be summarized in the following way. Sen suggests a 'midfare' theory, i.e. a theory that locates welfare between on the one hand the set of external resources, such as physical surroundings, money and social status, and on the other hand the mental result of utilizing those resources, such as pleasure and happiness. My own proposal, like most theories of welfare, locates welfare at the latter end and could be called a final-end theory.

An Outline of Sen's Theory of Welfare

I will here attempt to give an outline of Sen's theory of welfare. I will abstain, however, from following him in his more technical, economic, applications. Basically I will follow his presentation in *Inequality Reexamined* (1992) but will also quote a few of his other central writings, in particular *Commodities and Capabilities* (1985). It must be emphasized that Sen exclusively deals with human affairs. This means that all his examples are human. However, in a recent article on animal rights Martha Nussbaum (2004) has suggested how Sen's theory of capabilities could be transferred to the animal arena.

Sen reacts to two tendencies in modern (and indeed also ancient) theorizing on welfare. One tendency is the classic utilitarian, which identifies welfare with certain end-states of actions, such as pleasure or happiness. A theory of this kind has, according to Sen, the serious disadvantage that it can equal the welfare of the content beggar with the welfare of the billionaire. Such a result is absurd, he contends, if we wish to build a theory of welfare that could be of any use in, for instance, political theory.

Another tendency goes in the completely opposite direction. Theorists in this quarter attempt to identify people's welfare with the amount of valuable goods or resources that these people have. A famous theory in this category is that of John Rawls (1971), who identifies a set of *primary goods*, the possession of which determines the owner's state of goodness. (Rawls does not use the term 'welfare' or 'well-being'.) Among these goods are certain social ones, such as rights and liberties, whilst others are natural goods, including health, wealth and income.

Such a theory, says Sen, is also inadequate, since it does not take into account people's varying abilities to take advantage of goods. A person who cannot read cannot use all the brilliant books that he or she may possess. A person who is completely uneducated cannot take advantage of all the modern technology that modern Western society provides. Thus, the goods themselves cannot constitute welfare, according to Sen. Hence the kernel of welfare lies in the usage of goods, not in the goods themselves.

Sen's Theory of Functioning and Capabilities

The welfare of a person, Sen says, can be seen in terms of the quality of the person's living. Living, in its turn, consists of a set of interrelated functionings, including beings and doings.

> The relevant functionings can vary from such elementary things as being adequately nourished, being in good health, avoiding escapable morbidity and premature mortality, etc., to more complex achievements such as being happy, having self-respect, taking part in the life of the community, and so on.
>
> (Sen, 1992, p. 39)

An evaluation of a person's welfare must, according to Sen, take the form of an assessment of the constitutive elements of living, viz. the person's functionings.

In addition to functionings Sen mentions a further dimension which is highly relevant for a person's welfare. This is the capability-dimension of a person. Capabilities are crucial to a person's welfare, Sen thinks, because they constitute the person's real freedom to have welfare. Persons must, for instance, have a capability to take part in politics in order to achieve their political goals. But Sen also holds that the capability is a genuine part of welfare. 'Choosing may itself be a valuable part of living, and a life of genuine choice with serious options may be seen – for that reason – richer. In this view, at least some types of capabilities contribute *directly* to well-being, making one's life richer with the opportunity of reflective choice' (1992, p. 41).

The fundamental question now is: what capabilities and functionings are valuable from a welfare point of view? In answering this question Sen is surprisingly silent. What he has provided are a few indirect, and mostly negative, clues. Sen observes that not everything that a person wants and chooses to do is included in his or her welfare. A person's choice to drink alcohol, for instance, can hardly be seen as a part of this person's welfare. However,

> a person's goals will, in the case of most 'normal' people, include *inter alia* the pursuit of their own well-being. Indeed, the overall balance of agency objectives might be seen, with some plausibility, as reflecting the weights that the person herself would attach to her own well-being *among* the things that she wishes to promote.
>
> (Sen, 1992, p. 69)

Sen, then, seems to contend that we can for the most part follow the person's wishes in our assessment of the welfare of the person. On the other hand, Sen readily admits that a goal of a non-welfare kind may prevent the fulfilment of a welfare goal. Sen also notices the interesting case where the expansion of one's freedom in itself makes life more difficult to live. It provides the person with greater options. He or she, however, is disturbed by all these options. They may become a burden and prevent the living of a good life.

What Exactly are Functionings and Capabilities?

I will now go a little deeper into Sen's theory in order to give a preliminary assessment and compare it with my own ideas on health and welfare. Consider the following theoretical presentation in Sen (1985):

> In getting an idea of the well-being of the person, we clearly have to move on to 'functionings', to wit, what the person succeeds in *doing* with the commodities and characteristics at his or her command. For example, we must take note that a disabled person may not be able to do many things an able-bodied individual can, with the same bundle of commodities. A functioning is an achievement of a person: what he or she manages to do or to be. It reflects, as it were, a part of the 'state' of that person. It has to be distinguished from the commodities that are used to achieve those functionings. For example: bicycling has to be distinguished from possessing a bike. It has to be distinguished also from the happiness generated by the functioning, for example, actually cycling around must not be identified with the pleasure obtained from the act. A functioning is thus different both from (1) having goods (and the corresponding characteristics), to which it is posterior, and (2) having utility (in the form of happiness resulting from that functioning), to which it is, in an important way, prior.
>
> (Sen, 1985, pp. 10–11)

Again the general idea is quite clear. We must distinguish functionings from both external goods and the end states of actions, such as happiness and desire fulfilment. However, when one tries to get a deeper understanding of functionings difficulties arise. There is a tension in Sen's writings between a very wide interpretation of functionings, including practically all kinds of beings and doings of an individual, and quite a narrow interpretation, where there is a focus on the subject's achievements. The quotation just given represents the narrow concept. Sen's later work indicates something wider. Here he mentions being well-nourished and being in good health as examples of functionings. Neither of these states has to be an achievement on the part of its bearers.

In my preliminary reconstruction of Sen's theory I will follow the wider interpretation, supported in Sen (1993). Functionings are both doings and beings. All kinds of actions are functionings. So are most results of actions as long as they pertain to the subject (excluding, however, mental results, such as happiness). Being in a new place and being in a new relationship are functionings. It seems reasonable to include all kinds of properties among functionings.

These observations have important repercussions for the notion of capability. Capability is much more than ability in an action theoretic sense. It also covers opportunity. Sen (1993, p. 44) says: 'A person's ability to achieve various valuable functionings may be greatly enhanced by public action and policy,' Thus, capability is expanded by external policy, for instance by giving people a wider and more suitable platform for action. But this is not all. Capability also stands for such preconditions for functionings as have nothing to do with intentional action. External and internal conditions for becoming tall, for sleeping well, for having a healthy body, are all capabilities in Sen's broad sense.

To put matters in a simplified way: a functioning is any property (apart from logical necessities) of a person and a capability is any precondition for having such a property. From this we can draw some conclusions. Since all properties are at the same time preconditions for having a large set of other properties, all functionings are also capabilities for other functionings. Being well-nourished, which is a functioning, is a precondition for working well, which is another functioning. However, not all capabilities are functionings. This is so since not all capabilities are properties of the person in question. The configuration of states external to a person are capabilities. They are not, however, per definition, functionings of the person.

A Comparison between Sen's Theory and My Own Analysis of the Notions of Health and Welfare

My primary aim above has been to characterize the essential concepts in Sen's theory of welfare, viz. functionings and capabilities. I will now make some comparisons between Sen's theory and my own proposals in the area. Consider first Sen's theory in relation to my own analysis of health.

First, Sen's theory of welfare covers much more ground than my theory of health. This is an expected and highly reasonable result. Welfare must be a much broader concept than health. But, second, some of the constituents of Sen's theory of welfare are identical with the main constituents in my theory of health. A person's abilities to achieve vital goals are certainly important capabilities in Sen's theory. This result reflects Sen's general aim of identifying welfare on a person's 'midfare' level. Sen has explicitly mentioned health among his capabilities and functionings. Probably, however, he did not regard health so much from the capability perspective as I do.

Sen's theory of welfare is in a much clearer sense a rival to the kind of preference theory of happiness that I have been advocating earlier in this book. Moreover, Sen has explicitly taken issue with the latter kind of theory. Let me here scrutinize a central argument in Sen's criticism:

> A person who is ill-fed, undernourished, and ill can still be high up in the scale of happiness or desire-fulfilment if he or she has learned to have 'realistic' desires and to take pleasure in small mercies. The physical conditions of a person do not enter the view of well-being seen entirely in terms of happiness or desire fulfilment, except insofar as they are indirectly covered by the mental attitudes of happiness or desire.
>
> (Sen, 1985, p. 21)

This quotation reveals that Sen is a welfare theorist who has a social science platform rather than an individualistic or psychological platform. To Sen it seems *self-evident* that the person who is undernourished, unsheltered and ill, has a low degree of welfare. This position is quite similar to the position of some animal welfare theorists (see, for instance my discussion of Broom's theory, Chapter 10). I do not agree with it, however. And I will therefore articulate my view with regard to the quoted example.

First, the undernourished, unsheltered and ill person is in most instances a very unhappy person. Most persons suffer from being in such a state. Second, if the undernourished, unsheltered and ill person is not at present suffering there is, if the situation is grave, a high probability that he or she will suffer within a short time. Our empirical knowledge tells us that certain degrees of undernourishment, lack of shelter and illness must, unless treated, sooner or later lead to pain or other kinds of suffering. Thus, in my view, the undernourished etc. person is seriously *endangered* with respect to his or her welfare. There are very good reasons for taking action and helping this person. Hence, from a pragmatic point of view Sen and I would come to similar conclusions also in this case.

Third, consider the case of an undernourished etc. person who does not suffer, or for other reasons does not want to change his or her predicament, and whose situation is such that he or she might avoid suffering in the future. What is Sen's position concerning this person? If I understand him correctly we should still consider this person as severely lacking in welfare. The functionings of nourishment, being sheltered and having health are so important in themselves (having an absolute value?) that whatever the person's expressed wishes or lack of wishes, it would be absurd to assign a high degree of welfare to this person. I can, however, imagine circumstances when this person has a high degree of welfare. Thus, Sen and I part company on this essential point. I will consider some cases:

1. The conditions of undernourishment are not so severe. The subject is a strong person and can easily compensate for these conditions. He or she is doing well at work and is in general flourishing as a person. Thus, this person enjoys life as much as any ordinary person.

2. The subject is a man living in a poor country which is led by a despotic regime. Most of his fellow countrymen are indeed suffering. The man and some of his comrades have, however, found a successful way of resigning themselves to the external circumstances and have concentrated on developing some of their talents and interests which are still possible to cultivate in the country.

One may counter here that Sen has not had such examples in mind. If the persons involved grow as human beings, then there is a high degree of welfare also according to Sen. But if Sen were to counter in this way, then his theses about the undernourished etc. person do not have the absolute character that his wording seems to indicate.

One must remember that much work in health care and social care entails helping people to become realistic in order to cope with life. It is often difficult to change people's circumstances regarding economy, employment and housing. It is most of the time possible, however, to help people psychologically. One of the best means is to develop their attitudes to and expectations from life. By doing so one may contribute to a radical improvement of their happiness and thereby welfare. But Sen, given his theory, is forced to claim that their level of welfare cannot be raised simply by psychological means.

Another reason for Sen's critical attitude to a preference–satisfaction theory of welfare is that it entails a neglect of values. The welfare of a person is dependent on what this person values, he says, not just on what he or she desires. Here is a crucial point that has to be considered. (Consider my points about sustainable welfare above.) A person (as well as an animal) can have wants of quite different kinds and of different dignity. A man can want to scratch his nose, he can want to have a bottle of whisky and he can want to have a prestigious job. These wants differ with regard to the manner and degree in which their satisfaction contributes to the man's welfare. Some provide momentary pleasure, others may contribute to lasting satisfaction. The realization of some of the wants may in the long run indeed be detrimental.

If we are interested in the bearer's welfare in the long run we must analyse these wants carefully. We must consider the subject's hierarchy of wants and we must study whether the wants concern enduring states or quite temporary states of affairs. Only a few of these wants contribute much to the welfare of the subject. (For a discussion of complications here, see Nordenfelt, 1993, 2000.) On this point I think Sen and I agree.

I am more hesitant about Sen's method for selecting the crucial wants. He introduces the concept of *valuing* (1985, p. 32). He says that only the realization of what is valued by A can contribute to A's welfare. I agree that there are acts of valuation. Valuations influence our desires. So far Sen and I agree. The trouble, however, starts at this point. There are many kinds of valuation. There are moral valuations, aesthetic valuations, intellectual valuations, as well as valuations concerning welfare. Only the latter valuations are automatically relevant in our context. The satisfaction of a moral valuation may contribute to the subject's welfare if he or she is a well-integrated person and has incorporated these valuations among the personal wants. But, as Sen himself observes, choosing to act in accordance with one's ethics need not raise the level of one's welfare at all.

The reference to valuation in general cannot help us understand or define the criteria for welfare. We must know first which are the valuations of welfare. But for that we need a definition of welfare. Hence we are going in full circle.

Concluding Remarks

In this appendix I have attempted to relate my theories of health and welfare to Amartya Sen's much debated theory of welfare. I have found that health, as I understand it, is one element of Sen's functionings and capabilities that form his space of welfare. This is not a surprising conclusion. However, our theories of welfare differ. Sen is a midfare theorist of welfare. I am a final-end theorist. Sen has criticized the validity of my kind of theory. I have countered by questioning whether Sen can clarify his notion of welfare without referring to some kind of end-state, for instance happiness.

(This appendix summarizes an analysis presented in Nordenfelt, 2000.)

References

Aggernaes, A. (1994) On general and need-related quality of life. In: Nordenfelt, L. (ed.) *Concepts and Measurements of Quality of Life in Health Care*. Kluwer Academic Publishers, Dordrecht, The Netherlands, pp. 241–255.

Algers, B. (1990) Naturligt beteende – ett naturligt begrepp. *Svensk veterinärtidning* 42, 517–519.

Algers, B. and Jensen, P. (1985) Communication during suckling in the domestic pig. Effects of continuous noise. *Applied Animal Behaviour Science* 14, 49–61.

Alrøe, H.F., Vaarst, M. and Kristensen, E.S. (2001) Does organic farming face distinctive livestock welfare issues? A conceptual analysis. *Journal of Agricultural and Environmental Ethics* 14, 275–299.

Appleby, M.C. and Sandøe, P. (2002) Philosophical debate on the nature of well-being: implications for animal welfare. *Animal Welfare* 11, 283–294.

Aristotle (1934) *The Nicomachean Ethics*. Harvard University Press, The Loeb Classical Library, Cambridge, Massachusetts.

Aristotle (1982) *The Eudemian Ethics*, Books I, II and VIII, translated with a commentary by M. Woods. Clarendon Press, Oxford, UK.

Aristotle (1991) *History of Animals*, Books VII–X, edited and translated by D.M. Balme. Harvard University Press, Cambridge, Massachusetts.

Armstrong, D. (1968) *A Materialist Theory of the Mind*. Routledge & Kegan Paul, London.

Ayer, A.J. (ed.) (1978) *Logical Positivism*. P. Greenwood, Westport, Connecticut.

Barnard, C.J. and Hurst, J.L. (1996) Welfare by design: the natural selection of welfare criteria. *Animal Welfare* 5, 405–433.

Bennett, A.T. (1996) Do animals have cognitive maps? *The Journal of Experimental Biology* 199, 219–224.

Bentham, J. (1977) *A Fragment of Government*, the new authoritative edition by J.H. Burns and H.L.A. Hart. Cambridge University Press, Cambridge, UK.

Bentham, J. (1982) *An Introduction to the Principles of Morals and Legislation*, the new authoritative edition by J.H. Burns and H.L.A. Hart. Cambridge University Press, Cambridge, UK.

Bermond, B. (1997) The myth of animal suffering. In: Dol, M., Kasanmoentalib, S., Lijmbach, S., Rivas, E. and Van den Boss, R. (eds) *Animal Consciousness and Animal Ethics: Perspectives from The Netherlands*. Van Gorcum, Assen, The Netherlands, pp. 125–143.

Blood, D.C. and Studdert, V.P. (1999) *Saunders Comprehensive Veterinary Dictionary*. W.B. Saunders, London.

Boorse, C. (1977) Health as a scientific concept. *Philosophy of Science* 44, 542–573.

Boorse, C. (1997) Rebuttal on health. In: Humber, J. and Almeder, R. (eds) *What is Disease?* Humana Press, Totowa, New Jersey, pp. 1–134.

Brambell Committee (1965) *Report of the Technical Committee to Enquire into the Welfare of Animals kept under Intensive Livestock Husbandry Systems*, Command Report 2836. Her Majesty's Stationery Office, London.

Brooks, R. (1996) EuroQol: the current state of play. *Health Policy* 37, 53–72.

Broom, D.M. (1986) Indicators of poor welfare. *British Veterinary Journal* 142, 524–526.

Broom, D.M. (1988) The scientific assessment of animal welfare. *Applied Animal Behaviour Science* 20, 5–19.

Broom, D.M. (1991) Animal welfare: concepts and measurement. *Journal of Animal Science* 69, 4167–4175.

Broom, D.M. (1993a) A usable definition of animal welfare. *Journal of Agricultural and Environmental Ethics* 6, Suppl. 2, 15–25.

Broom, D.M. (1993b) Panel discussion. *Journal of Agricultural and Environmental Ethics* 6, Suppl. 2, 51–58.

Broom, D.M. (1996) Animal welfare defined in terms of attempts to cope with the environment. *Acta agriculturae Scandinavica. Section A, Animal Science. Supplementum* 27, 22–28.

Broom, D.M. (1998) Welfare, stress and the evolution of feelings. *Advances in the Study of Behavior* 27, 371–403.

Broom, D.M. (2001) Coping, stress and welfare. In: Broom, D.M. (ed.) *Coping with Challenge: Welfare in Animals Including Humans*. Dahlem University Press, Berlin, pp. 1–9.

Broom, D.M. and Johnson, K.G. (1993) *Stress and Animal Welfare*. Chapman & Hall, London.

Broom, D.M. and Kirkden, R.D. (2004) Welfare, stress, behaviour and pathophysiology. In: Dunlop, R.H. and Malbert, C.-H. (eds) *Veterinary Pathophysiology*. Blackwell, Ames, Iowa, pp. 337–369.

Brülde, B. (1998) *The Human Good*, Acta Philosophica Gothoburgensia 6. Gothenburg University, Gothenburg, Sweden.

Bshary, R., Wickler, W. and Fricke, H. (2002) Fish cognition: a primate's eye view. *Animal Cognition* 5, 1–13.

Burghardt, G.M. (1990) Animal suffering, critical anthropomorphism, and reproductive rights. *Behavioral and Brain Sciences* 13, 14–15.

Burkhardt, R.W. (1997) The founders of ethology and the problem of animal subjective experience. In: Dol, M., Kasanmoentalib, S., Lijmbach, S., Rivas, E. and Van den Boss, R. (eds) *Animal Consciousness and Animal Ethics: Perspectives from The Netherlands*. Van Gorcum, Assen, The Netherlands, pp. 1–13.

Burt, C. (1962) The concept of consciousness. *British Journal of Psychology* 52, 229–242.

Canguilhem, G. (1978) *On the Normal and the Pathological*. D. Reidel Publishing Company, Dordrecht, The Netherlands.

Cannon, W.B. (1932) *The Wisdom of the Body*. Norton, New York.

Cheville, N.F. (1988) *Introduction to Veterinary Pathology*. Iowa State University Press, Ames, Iowa.

Clark, J.D., Rager, D.R. and Calpin, J.P. (1997) Animal well-being I. General considerations. *Laboratory Animal Science* 47, 564–570.

Crisp, R. (1990) Evolution and psychological unity. In: Bekoff, M. and Jamieson, D. (eds) *Interpretation and Explanation in the Study of Behavior*, Vol. I. Westview Press, Boulder, Colorado, pp. 394–413.

Culver, C.M. and Gert, E. (1982) *Philosophy in Medicine: Conceptual and Ethical Issues in Medicine and Psychiatry*. Oxford University Press, Oxford, UK.

Curtis, S.E. (1987) Animal well-being and animal care. *Veterinary Clinics of North America: Food Animal Practice* 3, 369–381.

Darwin, C. (1871) *The Descent of Man*. John Murray, London.

Darwin, C. (1872) *The Expression of Emotions in Man and Animals*. John Murray, London.

Dawkins, M.S. (1980) *Animal Suffering: the Science of Animal Welfare*. Chapman and Hall, London.

Dawkins, M.S. (1983) Battery hens name their price: consumer demand theory and the measurement of ethological 'needs'. *Animal Behaviour* 31, 1195–1205.

Dawkins, M.S. (1988) Behavioural deprivation: a central problem in animal welfare. *Applied Animal Behaviour Science* 20, 209–225.

Dawkins, M.S. (1990) From an animal's point of view: motivation, fitness, and animal welfare. *Behavioral and Brain Sciences* 13, 1–9.

Day, C. (1995) *The Homeopathic Treatment of Beef & Dairy Cattle*. Beaconsfield Publishers Ltd, Beaconsfield, UK.

Descartes, R. (1992) *A Discourse on Method*. J.M. Dent & Sons Ltd, London.

Diagnostic and Statistical Manual of Mental Disorders (1994) The American Psychiatric Association, Washington DC.

Dolan, P., Gudex, C. and Kind, P. (1995) *A Social Tariff for EuroQol: Results from a UK General Population Survey*. Centre for Health Economics, York, UK.

Duncan, I.J.H. (1993) Welfare is to do with what animals feel. *Journal of Agricultural and Environmental Ethics* 6, Suppl. 2, 8–14.

Duncan, I.J.H. (1996) Animal welfare defined in terms of feelings. *Acta agriculturae Scandinavica. Section A, Animal Science. Supplementum* 27, 29–35.

Duncan, I.J.H. and Fraser, D. (1997) Understanding animal welfare. In: Appleby, M.C. and Hughes, B.O. (eds) *Animal Welfare*. CAB International, Wallingford, UK, pp. 19–31.

Dupré, J. (1990) The mental lives of nonhuman animals. In: Bekoff, M. and Jamieson, D. (eds) *Interpretation and Explanation in the Study of Behavior*, Vol. I. Westview Press, Boulder, Colorado, pp. 428–448.

Dupré, J. (2001) *Human Nature and the Limits of Science*. Clarendon Press, Oxford, UK.

Dupré, J. (2002) *Humans and Other Animals*. Clarendon Press, Oxford, UK.

Engelhardt, H.T. Jr (1986) *The Foundations of Bioethics*. Oxford University Press, Oxford, UK.

Feyerabend, P. (1975) *Against Method: Outline of an Anarchistic Theory of Knowledge*. Verso, London.

Fraser, A.F. and Broom, D.M. (1997) *Farm Animal Behaviour and Welfare*. Oxford University Press, Oxford, UK.

Fraser, D. (1995) Science values and animal welfare: exploring the 'inextricable connection'. *Animal Welfare* 4, 103–117.

Fraser, D. (1999) Animal ethics and animal welfare science: bridging the two cultures. *Applied Animal Behaviour Science* 65, 171–189.

Fraser, D. and Duncan, I.J.H. (1998) 'Pleasures', 'pains' and animal welfare: toward a natural history of affect. *Animal Welfare* 7, 383–396.

Fraser, D., Weary, D.M., Pajor, E.A. and Milligan, B.N. (1997) A scientific conception of animal welfare that reflects ethical concerns. *Animal Welfare* 6, 187–205.

Fulford, K.W.M. (1989) *Moral Theory and Medical Practice*. Cambridge University Press, Cambridge, UK.

Gadamer, H.-G. (1993) *Über die Verborgenheit der Gesundheit*. Suhrkamp Verlag, Frankfurt am Main, Germany.

Galen (1997) *Selected Works*, translated with an introduction and notes by P.N. Singer. Oxford University Press, Oxford, UK.

Goldie, P. (2000) *The Emotions*. Clarendon Press, Oxford, UK.

Gonyou, H.W. (1993) Animal welfare: definitions and assessment. *Journal of Agricultural and Environmental Ethics* 6, Suppl. 2, 37–43.

Gould, S.J. (2000) More things in heaven and earth. In: Rose, H. and Rose, S. (eds) *Alas, Poor Darwin: Arguments Against Evolutionary Psychology*. Jonathan Cape, London, pp. 85–105.

Gould, S.J. and Vrba, E.S. (1998) Exaptation: a missing term in the science of form. In:

Allen, C., Bekoff, M. and Lauder, G. (eds) *Nature's Purposes: Analyses of Function and Design in Biology.* The MIT Press, Boston, Massachusetts, pp. 519–540.

Griffin, D.R. (1984) *Animal Thinking.* Harvard University Press, Cambridge, Massachusetts.

Griffin, D.R. (2001) *Animal Minds: Beyond Cognition to Consciousness.* Chicago University Press, Chicago, Illinois.

Grunsell, C.S.G. (1995) Health. In: West, G.P. (ed.) *Black's Veterinary Dictionary.* A & C Black, London.

Gunnarsson, S. (2005) Definitions of Health and Disease in Textbooks of Veterinary Medicine (in press).

Häring, B. (1987) *Medical Ethics.* St Paul Publications, Slough, UK.

Harrison, R. (1964) *Animal Machines: The New Factory Farming Industry.* Stuart, London.

Harrison, R. (1983) *Bentham.* Routledge & Kegan Paul, London.

Hellström, O. (1993) The importance of a holistic concept of health for health care: examples from the clinic. *Theoretical Medicine* 14, 325–342.

Henriksson, M. and Carlsson, P. (2002) *Att mäta hälsorelaterad livskvalitet – en beskrivning av instrumentet EQ-5D*, CMT Report 2002:1. Center for Medical Technology Assessment, Linköping University, Linköping, Sweden.

Heyes, C.M. (1998) Theory of mind in non-human primates. *Behavioral and Brain Sciences* 21, 101–148.

Hovi, M., Gray, D., Vaarst, M., Striezel, A., Walkenhorst, M. and Roderick, S. (2004) Promoting health and welfare through planning. In: Vaarst, M., Roderick, S., Lund, V. and Lockeretz, W. (eds) *Animal Health and Welfare in Organic Agriculture.* CAB International, Wallingford, UK, pp. 253–277.

Hughes, B.O. (1980) The assessment of behavioural needs. In: Moss, R. (ed.) *The Laying Hen and its Environment.* Martinus Nijhoff, Boston, Massachusetts.

Hughes, B.O. and Curtis, P.E. (1997) Health and disease. In: Appleby, M.C. and Hughes, B.O. (eds) *Animal Welfare.* CAB International, Wallingford, UK, pp. 109–125.

Hume, D. (1961) *A Treatise of Human Nature*, Vol. 1. J.M. Dent & Sons Ltd, London.

Hunt, S.M. (1988) Subjective health indicators and health promotion. *Health Promotion* 3, 23–34.

Hurnik, J.F. and Lehman, H. (1985) The philosophy of farm animal welfare: a contribution to the assessment of farm animal well-being. In: Wegner, R.-M. (ed.) *Proceedings from the Second European Symposium on Poultry Welfare.* Federal Agricultural Research Centre, Braunschweig-Völkenrode, Germany, pp. 256–265.

Huxley, J.S. (1914) The courtship habits of the great crested grebe (*Podiceps cristatus*); with an addition to the theory of sexual selection. *Proceedings of the Zoological Society of London* 35, 491–562.

James, W. (1901) *The Principles of Psychology*, Vol. 1. MacMillan and Co. Ltd, London.

De Jonge, F.H. (1997) Animal welfare? An ethological contribution to the understanding of emotions in pigs. In: Dol, M., Kasanmoentalib, S., Lijmbach, S., Rivas, E. and Van den Boss, E. (eds) *Animal Consciousness and Animal Ethics: Perspectives from The Netherlands.* Van Gorcum, Assen, The Netherlands, pp. 103–113.

Kajandi, M. (1994) A psychiatric and interactional perspective on quality of life. In: Nordenfelt, L. (ed.) *Concepts and Measurement of Quality of Life in Health Care.* Kluwer Academic Publishers, Dordrecht, The Netherlands, pp. 257–276.

Kallenberg, K., Bråkenhielm, C.R. and Larsson, G. (1997) *Tro och värderingar i 90-talets Sverige [Beliefs and Evaluations in Present Day Sweden].* Libris, Örebro, Sweden.

Karavanich, C. and Atema, J. (1998) Individual recognition and memory in lobster dominance. *Animal Behaviour* 56, 1553–1560.

Kenny, A. (1963) *Action, Emotion and Will.* Routledge & Kegan Paul, London.

Khushf, G. (2001) What is at issue in the debate about concepts of health and disease? Framing the problem of demarcation for a post-positivist era of medicine. In: Nordenfelt, L. (ed.) *Health, Science, and Ordinary Language.* Rodopi Publishers, Amsterdam, pp. 123–169.

Kirk, S. and Kutchins, H. (1997) *Making Us Crazy: DSM, The Psychiatric Bible and the Creation of Mental Disorders.* Free Press, New York.

Knierim, U., Carter, C.S., Fraser, D., Gärtner, K., Lutgendorf, S.K., Mineka, S., Panksepp, J. and Sachser, N. (2001) Group Report: Good welfare: improving quality of life. In: Broom, D.M. (ed.) *Coping with Challenge: Welfare in Animals including Humans.* Dahlem University Press, Berlin, pp. 79–100.

Kuhn, T.S. (1962) *The Nature of Scientific Revolutions.* University of Chicago Press, Chicago, Illinois.

Leder, D. (1990) Clinical interpretation: the hermeneutics of medicine. *Theoretical Medicine* 11, 9–24.

Liss, P.-E. (1993) *Health Care Need.* Avebury, Ashgate, UK.

Lopez, F., Serrano, J.M. and Acosta, F.J. (1994) Parallels between the foraging strategies of ants and plants. *Tree* 9, 150–152.

Lorenz, K.Z. (1937) Über die Bildung des Instinktbegriffes. *Die Naturwissenschaften* 25, 289–331.

Lund, V. and Röcklingsberg, H. (2001) Outlining a conception of animal welfare for organic farming systems. *Journal of Agricultural and Environmental Ethics* 14, 391–424.

Luria, A.R. (1980) *Higher Cortical Functions in Man*, 2nd edn. Plenum Publishing Corporation, New York.

McGill, V.J. (1967) *The Idea of Happiness.* Fredrick A. Jaeger Publishers, New York.

McGlone, J. (1993) What is animal welfare? *Journal of Agricultural and Environmental Ethics* 6, Suppl. 2, 26–36.

McMillan, F.D. (2000) Quality of life in animals. *Journal of the American Veterinary Medical Association* 216, 1904–1910.

McMillan, F.D. (2003) A world of hurts – is pain special? *Journal of the American Veterinary Medical Association* 223, 183–186.

McMillan, F.D. and Rollin, B.E. (2001) The presence of mind: on reunifying the animal mind and body. *Journal of the American Veterinary Medical Association* 218, 1723–1727.

Maslow, A. (1968) *Toward a Psychology of Being*, 2nd edn. D. van Nostrand, New York.

Mayr, E. (1982) *The Growth of Biological Thought: Diversity, Evolution and Inheritance.* Harvard University Press, Cambridge, Massachusetts.

Moum, T. (1994) Needs, rights and resources in quality of life research. In: Nordenfelt, L. (ed.) *Concepts and Measurement of Quality of Life in Health Care.* Kluwer Academic Publishers, Dordrecht, The Netherlands, pp. 79–93.

Naess, S. (1987) *Quality of Life Research.* Institute of Applied Social Research, Oslo.

Ng, Y.-K. (1995) Towards welfare biology: evolutionary economics of animal consciousness and suffering. *Biology and Philosophy* 10, 255–285.

Nordenfelt, L. (1987) *On the Nature of Health.* D. Reidel Publishing Company, Dordrecht, The Netherlands.

Nordenfelt, L. (1993) *Quality of Life, Health and Happiness.* Avebury, Aldershot, UK.

Nordenfelt, L. (ed.) (1994) *Concepts and Measurement of Quality of Life in Health Care.* Kluwer Academic Publishers, Dordrecht, The Netherlands.

Nordenfelt, L. (1995) *On the Nature of Health*, 2nd, revised, edn. Kluwer Academic Publishers, Dordrecht, The Netherlands.

Nordenfelt, L. (2000) *Action, Ability and Health.* Kluwer Academic Publishers, Dordrecht, The Netherlands.

Nordenfelt, L. (2001) *Health, Science, and Ordinary Language.* Rodopi Publishers, Amsterdam.

Nordenfelt, L. (2003) Health as natural function. In: Nordenfelt, L. and Liss, P.-E. (eds)

Dimensions of Health and Health Promotion. Rodopi Publishers, Amsterdam, pp. 37–54.

Nussbaum, M.C. (2004) Beyond 'compassion and humanity': justice for non-human animals. In: Sunstein, C.R. and Nussbaum, M.C. (eds) Animal Rights: Current Debates and New Directions. Oxford University Press, Oxford, UK.

Nussbaum, M.C. and Sen, A. (eds) (1993) The Quality of Life. Clarendon Press, Oxford, UK.

Panksepp, J. (1998) Affective Neuroscience: The Foundations of Human and Animal Emotions. Oxford University Press, Oxford, UK.

Parsons, T. (1972) Definitions of health and illness in the light of American values and social structure. In: Jaco, E.G. (ed.) Patients, Physicians, and Illness. The Free Press, New York.

Pellegrino, E.D. and Thomasma, D.C. (1981) A Philosophical Basis of Medical Practice. Oxford University Press, Oxford, UK.

Plato (1998) The Republic. In: Cooper, J.M. (ed.) Complete Works of Plato. Hackett Publishing Company, Indianapolis, Indiana.

Pörn, I. (1986) On the nature of emotions. In: Needham, P. and Odelstad, J. (eds) Changing Positions, Philosophical Studies No. 38. University of Uppsala, Uppsala, Sweden.

Pörn, I. (1993) Health and adaptedness. Theoretical Medicine 14, 295–304.

Rawls, J. (1971) A Theory of Justice. Harvard University Press, Boston, Massachusetts.

Reznek, L. (1987) The Nature of Disease. Routledge & Kegan Paul, London.

Richards, J.G. and Mohler, H. (1984) Benzodiazepine receptors. Neuropharmacology 23, 233–242.

Rollin, B.E. (1990a) How the animals lost their minds: animal mentation and scientific ideology. In: Bekoff, M. and Jamieson, D. (eds) Interpretation and Explanation in the Study of Behavior, Vol. I. Westview Press, Boulder, Colorado, pp. 375–393.

Rollin, B.E. (1990b) The Unheeded Cry: Animal Consciousness, Animal Pain and Science. Oxford University Press, Oxford, UK.

Rollin, B.E. (1992) Animal Rights and Human Morality. Prometheus Books, Buffalo, New York.

Rollin, B.E. (1996) Ideology, 'value-free science', and animal welfare. Acta agriculturae Scandinavica. Section A, Animal Science. Supplementum 27, 5–10.

Romanes, G.J. (1884, reprinted 1969) Mental Evolution in Animals. AMS Press, New York.

Rowan, A.N. (1988) Animal anxiety and animal suffering. Applied Animal Behaviour Science 20, 135–142.

Ryle, G. (1949) The Concept of Mind. Hutchinson, London.

Sandøe, P. (1996) Animal and human welfare – are they the same kind of thing? Acta agriculturae Scandinavica. Section A, Animal Science. Supplementum 27, 11–15.

Sandøe, P. and Kappell, D. (1994) Changing preferences: conceptual problems in comparing health-related quality of life. In: Nordenfelt, L. (ed.) Concepts and Measurement of Quality of Life in Health Care. Kluwer Academic Publishers, Dordrecht, The Netherlands, pp. 161–180.

Seedhouse, D. (1986) Health: Foundations of Achievement. John Wiley & Sons, Chichester, UK.

Sen, A. (1985) Commodities and Capabilities. North Holland, Amsterdam.

Sen, A. (1992) Inequality Reexamined. Clarendon Press, Oxford, UK.

Sen, A. (1993) Capability and well-being. In: Nussbaum, M.C. and Sen, A. (eds) The Quality of Life. Clarendon Press, Oxford, UK, pp. 30–53.

Simonsen, H.B. (1996) Assessment of animal welfare by a holistic approach: behaviour, health and measured opinion. Acta agriculturae Scandinavica. Section A, Animal Science. Supplementum 27, 91–96.

Singer, P. (1990) The significance of animal suffering. Behavioral and Brain Sciences 13, 9–12.

Singhal, G.D. and Patterson, T.J.S. (1993)

Synopsis of Ayurveda, based on a translation of the Treatise of Susruta. Oxford University Press, Oxford, UK.

Spinka, M., Newberry, R.C. and Bekoff, M. (2001) Mammalian play: training for the unexpected. *The Quarterly Review of Biology* 76, 141–168.

Sundström, P. (1987) *Icons of Disease*, Linköping Studies in Arts and Science, Vol. 14. University of Linköping, Linköping, Sweden.

Svenaeus, F. (2001) *The Hermeneutics of Medicine and the Phenomenology of Health*. Kluwer Academic Publishers, Dordrecht, The Netherlands.

Szasz, T. (1974) *The Myth of Mental Illness*, revised edn. Harper & Row, New York.

Tannenbaum, J. (1991) Ethics and animal welfare: the inextricable connection. *Journal of the American Veterinary Medical Association* 198, 1360–1376.

Tatarkiewicz, W. (1976) *Analysis of Happiness*. Martinus Nijhoff, The Hague, The Netherlands.

Taylor, P.W. (1986) *Respect for Nature: A Theory of Environmental Ethics*. Princeton University Press, Princeton, New Jersey.

Telfer, E. (1980) *Happiness*. MacMillan Press, London.

Temkin, O. (1963) The scientific approach to disease: specific entity and individual sickness. In: Crombie, A.C. (ed.) *Scientific Change: Historical Studies in the Intellectual, Social and Technical Conditions for Scientific Discovery and Technical Invention from Antiquity to the Present*. Basic Books, New York, pp. 629–647.

Toates, F. (1987) The relevance of models of motivation and learning to animal welfare. In: Wiepkema, P.R. and van Adrichen, P.W.M. (eds) *Biology of Stress in Farm Animals: An Integrative Approach*. Martinus Nijhoff Publishers, Dordrecht, The Netherlands, pp. 153–186.

Trigg, R. (1970) *Pain and Emotion*. Clarendon Press, Oxford, UK.

Twaddle, A.R. (1993) Disease, illness and sickness revisited. In: Twaddle, A. and Nordenfelt, L. (eds) *Disease, Illness and Sickness: Three Central Concepts in the Theory of Health*, Studies on Health and Society No. 18. Linköping University, Linköping, Sweden.

Veenhoven, R. (1984) *Conditions of Happiness*. D. Reidel Publishing Company, Dordrecht, The Netherlands.

Veenhoven, R. (2000) The four qualities of life: ordering concepts and measures of the good life. *Journal of Happiness Studies* 1, 1–39.

Vorstenbosch, J. (1997) Conscientiousness and consciousness. How to make up our minds about the animal mind. In: Dol, M., Kasanmoentalib, S., Lijmbach, S., Rivas, E. and Van den Boss, R. (eds) *Animal Consciousness and Animal Ethics: Perspectives from The Netherlands*. Van Gorcum, Assen, The Netherlands, pp. 32–47.

Wakefield, J.C. (1992a) The concept of mental disorder: on the boundary between biological facts and social values. *American Psychologist* 47, 373–388.

Wakefield, J.C. (1992b) Disorder as harmful dysfunction: a conceptual critique of DSM-III-R's definition of mental disorder. *Psychological Review* 99, 232–247.

Watson, J.B. (1925) *Behaviorism*. W.W. Norton, New York.

Webster, J. (1987) *Understanding the Dairy Cow*. BSP Professional Books, Oxford, UK.

Webster, J. (1994) *Animal Welfare: A Cool Eye Towards Eden*. Blackwell Science, Oxford, UK.

Whitbeck, C. (1981) A theory of health. In Caplan, A.L., Engelhardt, H.T. Jr and McCartney, J.J. (eds) *Concepts of Health and Disease: Interdisciplinary Perspectives*. Addison-Wesley Publishing Company, Reading, Massachusetts.

Wiepkema, P.R. (1985) Biology of fear. In: Wegner, R.-M. (ed.) *Proceedings from the Second European Symposium on Poultry Welfare*. Federal Agricultural Research Centre, Braunschweig-Völkenrode, Germany, pp. 84–93.

Wiepkema, P.R. (1987) Behavioural aspects of stress. In: Wiepkema, P.R. and van Adrichem, P.W.M. (eds) *Biology of Stress in Farm Animals: an Integrative Approach*. Martinus Nijhoff Publishers, Dordrecht, The Netherlands, pp. 113–133.

Wiepkema, P.R., Schouten, W.G.P. and Koene, P. (1993) Biological aspects of animal welfare: new perspectives. *Journal of Agricultural and Environmental Ethics* 6 (Suppl. 2), 92–100.

Williams, A. (1995) *The Measurement and Valuation of Health: a Chronicle*. Centre for Health Economics, York, UK.

Williams, G.C. (1966) *Adaptation and Natural Selection: a Critique of Some Current Evolutionary Thought*. Princeton University Press, Princeton, New Jersey.

Wittgenstein, L. (1953) *Philosophical Investigations*. Routledge & Kegan Paul, London.

World Health Organization (1948) *Official Records of the World Health Organization*, Vol. 2. WHO, Geneva, Switzerland, p. 100.

Von Wright, G.H. (1963) *The Varieties of Goodness*. Routledge & Kegan Paul, London.

Von Wright, G.H. (1995) Om behov [On needs]. In: Klockars, K. and Österman, B. (eds) *Begrepp om hälsa [Concepts of Health]*. Liber, Stockholm, Sweden, pp. 43–59.

Index